NFPA's
Illustrated Dictionary of
Fire Service Terms

JONES AND BARTLETT PUBLISHERS
Sudbury, Massachusetts
BOSTON   TORONTO   LONDON   SINGAPORE

**Jones and Bartlett Publishers
World Headquarters**
40 Tall Pine Drive
Sudbury, MA 01776
978-443-5000
info@jbpub.com
www.jbpub.com

Jones and Bartlett Publishers Canada
6339 Ormindale Way
Mississauga, Ontario L5V 1J2
Canada

Jones and Bartlett Publishers International
Barb House, Barb Mews
London W6 7PA
United Kingdom

National Fire Protection Association
1 Batterymarch Park
Quincy, MA 02169-7471
www.NFPA.org

Jones and Bartlett's books and products are available through most bookstores and online booksellers. To contact Jones and Bartlett Publishers directly, call 800-832-0034, fax 978-443-8000, or visit our website www.jbpub.com.

Substantial discounts on bulk quantities of Jones and Bartlett's publications are available to corporations, professional associations, and other qualified organizations. For details and specific discount information, contact the special sales department at Jones and Bartlett via the above contact information or send an email to specialsales@jbpub.com.

Production Credits
Chief Executive Officer: Clayton E. Jones
Chief Operating Officer: Donald W. Jones, Jr.
President, Higher Education and
 Professional Publishing: Robert W. Holland, Jr.
V.P., Sales and Marketing: William J. Kane
V.P., Production and Design: Anne Spencer
V.P., Manufacturing and Inventory Control: Therese Connell
Publisher, Public Safety Group: Kimberly Brophy
Acquisition Editor: William Larkin

Editor: Jennifer L. Reed
Production Editor: Jenny L. McIsaac
Director of Marketing: Alisha Weisman
Interior Design: Anne Spencer
Cover Design: Kristin E. Ohlin
Photo Research Manager/Photographer: Kimberly Potvin
Composition: Auburn Associates, Inc.
Text Printing and Binding: Malloy, Inc.
Cover Printing: Malloy, Inc.

Notice Concerning Liability: Publication of this work is for the purpose of circulating information and opinion among those concerned for fire and life safety and related subjects. While every effort has been made to achieve a work of high quality, neither the Publisher nor the NFPA guarantee the accuracy or assume any liability in connection with the information and opinions contained in this work. The Publisher and the NFPA shall in no event be liable for personal injury, property, or other damages of any nature whatsoever, whether special, indirect, consequential, or compensatory, directly or indirectly resulting from the publication, use of or reliance upon this work.

This work is published with the understanding that the Publisher and the NFPA are supplying information and opinion but are not attempting to render engineering or other professional services. If such services are required, the assistance of an appropriate professional should be sought.

Copyright 2006 by Jones and Bartlett Publishers, Inc. and the National Fire Protection Association.

All rights reserved. No part of the material protected by this copyright notice may be reproduced or utilized in any form, electronic or mechanical, including photocopying, recording, or by any information storage and retrieval system, without written permission from the copyright owner.

Additional credits appear on page 189, which constitutes a continuation of the copyright page.

ISBN-13: 978-0-7637-3909-6

Library of Congress Cataloging-in-Publication Data
NFPA's illustrated dictionary of fire service terms / National Fire Protection Association.
 p. cm.

 ISBN 0-7637-3909-X (pbk.)

 1. Fire extinction—Dictionaries. 2. Fire prevention—Dictionaries. I. Title: Illustrated dictionary of fire service terms. II. National Fire Protection Association.
 TH9116.N475 2006
 628.9'203—dc22
6048

2005032045

Printed in the United States of America
10 09 08 07 06   10 9 8 7 6 5 4 3 2 1

Preface

THE PURPOSE of the NFPA's *Illustrated Dictionary of Fire Service Terms* is to provide a consistent set of terminology as used throughout the *National Fire Codes®*. While to some extent the dictionary serves the purpose of an NFPA technical glossary, unlike most dictionaries, it has not been written by a group of professional lexicographers. Instead, it reflects the consensus viewpoints of the approximately 300 documents administered by NFPA's technical committees, each approaching its work from the viewpoint of its own technical field or discipline. The dictionary includes only the terms that appear in the *National Fire Codes®*. The code from which the definition has been taken is indicated at the end of each definition. Terms related to fire department standards are not included.

The definitions were written by people active in a particular field and reflect the sense in which that field uses the term defined. Comparison of definitions for the same term originating from different documents can reveal subtle, and sometimes substantial, differences in the concept represented by these terms in different fields. As a result, the redundant terms have been deleted from this edition.

Each definition is not merely the work of a small group. Instead, it is the product of the same rigorous consensus process through which all NFPA documents must pass before they are approved.

In some cases two or more definitions for the same term differ only slightly in language but not at all in meaning. This has led to an undesirable proliferation of definitions. In such cases, editorial and technical staff have evaluated these definitions and included only the definition that best describes the term for inclusion in this dictionary. Efforts by all NFPA technical committees and their respective staff liaisons are on-going to reduce the number of these redundancies, but additional efforts are needed. Accordingly, this edition of the dictionary should be considered a work in progress.

Numerous illustrations have been included to add clarification of the technical concepts defined.

Acknowledgments

A project of this type is never the result of a single person's effort, and this book is no exception. Both David Hague and Nancy Walker are deserving of exceptional credit for accurately keying literally thousands of definitions and maintaining a usable database. Dana Richards was instrumental in putting the final touches on the manuscript. This book would not exist if it were not for their efforts.

Aboveground Storage Tank. A horizontal or vertical tank that is listed and intended for fixed installation, without backfill, above or below grade, and is used within the scope of its approval or listing. NFPA 30A, 2003 ed.

Abrasion. The damaging effect on rope and other equipment caused by friction-like movement. NFPA 1670, 1999 ed.

Absolute Pressure. Pressure based on a zero reference point, the perfect vacuum. *Note: Measured from the zero reference point, the standard atmospheric pressure at sea level is an absolute pressure of 101.325 kPa (14.7 psia). Absolute pressure in the inch-pound system is commonly denoted in terms of pounds per square inch absolute (psia).* NFPA 1, 2003 ed.

Absolute Temperature. A temperature measured in Kelvins (K) or Rankines (R). NFPA 921, 2004 ed.

Absorption. The process in which materials hold liquids through the process of wetting. NFPA 471, 2002 ed.

Abuse. Harmful behaviors and/or actions, as defined by local law, that place an individual at risk and require reporting. NFPA 1035, 2005 ed.

Accelerant. A fuel or oxidizer, often an ignitable liquid, used to initiate a fire or increase the rate of growth or spread of fire. NFPA 921, 2004 ed.

Acceptable Entry Conditions. Conditions that must exist in a space to allow entry and to ensure that employees can safely enter into and work within the space. NFPA 1670, 2004 ed.

Acceptance. An agreement between the purchasing authority and the contractor that the terms and conditions of the contract have been met. NFPA 1906, 2001 ed.

Acceptance Tests. *As applies to fire apparatus:* Tests performed on behalf of or by the purchaser at the time of delivery to determine compliance with the specifications for the fire apparatus. NFPA 1901, 2003 ed. *As applies to marine fire fighting apparatus:* Tests performed on behalf of the purchaser by the manufacturer's representative at the time of delivery to determine compliance to the authority having jurisdiction requirements. NFPA 1925, 2004 ed.

Accepted Engineering Practices. Those requirements that are compatible with standards of practice required by a registered professional engineer. NFPA 1670, 1999 ed.

Access Box. An approved secure box, accessible by the authority having jurisdiction's master key or control, containing entrance keys or other devices to gain access to a structure or area. NFPA 1, 2003 ed.

Accessories. Those items that are attached to a proximity protective ensemble element but designed in such a manner to be removable from the proximity protective ensemble element and that are not necessary to meet the requirements of NFPA 1971. NFPA 1971, 2000 ed.

Accessory. An item that is attached to the certified product that is not necessary to meet the requirements of NFPA 1981. NFPA 1981, 2002 ed.

Accessory Structure. Any structure used incidentally to another structure. NFPA 1144, 2003 ed.

Accident. An unplanned event that interrupts an activity and sometimes causes injury or damage or a chance occurrence arising from unknown causes; an unexpected happening due to carelessness, ignorance, and the like. NFPA 921, 2004 ed.

Accommodation Spaces. Spaces designed for human occupancy as living spaces for persons aboard a vessel. NFPA 1925, 2004 ed.

Accountability. A system or process to track resources at an incident scene. NFPA 1561, 2005 ed.

Accredit. To give official authorization to or to approve a process or procedure, to recognize as conforming to a standard, and to recognize an entity (e.g., an educational institution) as maintaining standards that qualify its graduates for admission to higher or more specialized institutions or for professional practice. NFPA 1000, 2000 ed.

Accrediting Body. A voluntary, nongovernmental association that administers accrediting procedures for entities that certify individuals to fire service professional qualifications standards. NFPA 1000, 2000 ed.

Acoustic Emission Testing. A method of nondestructive testing (NDT) that utilizes acoustic or sound waves. NFPA 1914, 2002 ed.

Acquired Building. A structure acquired by the authority having jurisdiction from a property owner for the purpose of conducting live fire training evolutions. NFPA 1403, 2002 ed.

Acquired Prop. A piece of equipment such as an automobile that was not designed for burning but is used for live fire training evolutions. NFPA 1403, 2002 ed.

Activation Energy. The minimum energy that colliding fuel and oxygen molecules must possess to permit chemical interaction. NFPA 53, 2004 ed.

Activation Interval. A measurement that begins when the response unit is first notified of an incident and ends at the time that unit begins movement toward the incident. NFPA 450, 2004 ed.

Active Horizontal Angles of Light Emission. The angles, measured in a horizontal plane passing through the optical center of the optical source, as specified by the manufacturer of the optical device, between which the optical source contributes optical power. NFPA 1901, 2003 ed.

Active Search Measures. The phase of search measures that includes those that are formalized and coordinated with other agencies. NFPA 1006, 2003 ed.

Activity. A component of a public fire and life safety education program. NFPA 1035, 2005 ed.

Actual Response Time. The total period of time measured from the time of an alarm until the first ARFF vehicle arrives at the scene of an aircraft accident and is in position to apply agent. NFPA 403, 2003 ed.

Adapter. Any device that allows fire hose couplings to be safely interconnected with couplings of different sizes, threads, or mating surfaces, or that allows fire hose couplings to be safely connected to other appliances. NFPA 1963, 1998 ed.

Addition. An increase in building area, aggregate floor area, height, or number of stories of a structure. NFPA 5000, 2002 ed.

Address. A number or other code and the street name identifying a location. NFPA 450, 2004 ed.

Addressable Public Alerting System (APAS). A system that transmits a specific alert to a specific location or public alerting appliance or to multiple specific locations or appliances. NFPA 1221, 2002 ed.

Adequate and Reliable Water Supply. A water supply that is sufficient every day of the year to control and extinguish anticipated fires in the municipality, particular building, or building group served by the water supply. NFPA 1142, 2001 ed.

Adjust. To maintain or regulate, within prescribed limits, by setting the operating characteristics to specified parameters. NFPA 1915, 2000 ed.

Adjusting Device. An auxiliary equipment system component; a connector device that allows adjustment to be made to a piece of equipment. NFPA 1983, 2001 ed.

Admiralty Law/Maritime Law. A court exercising jurisdiction over maritime cases. NFPA 1405, 2001 ed.

Adsorption. The process in which a sorbate (hazardous liquid) interacts with a solid sorbent surface. NFPA 471, 2002 ed.

Advanced Cardiac Life Support (ACLS). A nationally recognized curriculum to teach advanced methods of treatment for cardiac and other emergencies. NFPA 450, 2004 ed.

Advanced Cleaning. The thorough cleaning of ensembles or elements by washing with cleaning agents. NFPA 1851, 2001 ed.

Advanced Life Support (ALS). Emergency medical treatment beyond basic life support level as defined by the medical authority having jurisdiction. NFPA 1500, 2002 ed.

Advanced Life Support (ALS) (EMS). Functional provision of advanced airway management including intubation, advanced cardiac monitoring, manual defibrillation, establishment and maintenance of intravenous access, and drug therapy. (See Figure A.1.) NFPA 1720, 2004 ed.

Aerial Device. An aerial ladder, elevating platform, or water tower that is designed to position personnel, handle materials, provide continuous egress, or discharge water. NFPA 1901, 2003 ed.

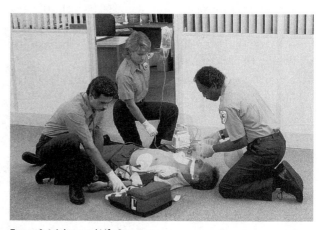

FIGURE A.1 Advanced Life Support.

Aerial Fire Apparatus. A vehicle equipped with an aerial ladder, elevating platform, aerial ladder platform, or water tower that is designed and equipped to support fire fighting and rescue operations by positioning personnel, handling materials, providing continuous egress, or discharging water at positions elevated from the ground. (See Figure A.2.) NFPA 1901, 2003 ed.

Aerial Ladder. A self-supporting, turntable-mounted, power-operated ladder of two or more sections permanently attached to a self-propelled automotive fire apparatus and designed to provide a continuous egress route from an elevated position to the ground. NFPA 1901, 2003 ed.

Aerial Ladder Platform. A type of aerial device that combines an elevating platform with the continuous egress capabilities of an aerial ladder. NFPA 1901, 1999 ed.

Aerial Ladder Sections. The structural members of the aerial ladder consisting of the base and fly sections. NFPA 1914, 2002 ed.

Aerial Operator. The fire apparatus driver who has met the requirements of Chapter 6 of NFPA 1002 for the operation of apparatus equipped with aerial devices. NFPA 1002, 2003 ed.

Aerosol. A product that is dispensed from an aerosol container by a propellant. (See Figure A.3.) NFPA 30B, 2002 ed.

After (Aft). The direction toward the stern of the vessel. NFPA 1405, 2001 ed.

Afterflame Time. The length of time for which a material, component, or chemical-protective suit continues to burn after the simulated chemical flash fire has ended. NFPA 1991, 2005 ed.

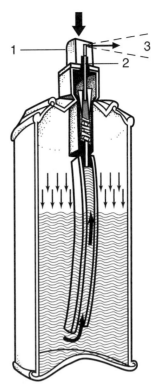

FIGURE A.3 Aerosol Container.

Agency Representative. An individual assigned to an incident from an assisting or cooperating agency who reports to the incident liaison officer and who has been delegated authority to make decisions on matters affecting that agency's participation at the incident. NFPA 1561, 2005 ed.

Aggressive Tire Tread. Tread designed to provide maximum traction for all types of surfaces, including sand, mud, snow, ice, and hard surfaces, wet or dry. NFPA 414, 2001 ed.

Air Accident Investigations Branch (AAIB). A UK agency that is responsible for investigating and determining the probable cause of all British aircraft accidents. NFPA 402, 2002 ed.

Air Control Panel. A consolidated arrangement of valves, regulators, gauges, and air system piping at a location that allows the operator to monitor and control the airflow and pressure within the air system from a centralized location. NFPA 1901, 2003 ed.

Air Quality Monitors. Instruments that monitor the air for such elements as carbon monoxide levels, moisture levels, and percent of oxygen. NFPA 1901, 2003 ed.

FIGURE A.2 Aerial Fire Apparatus.

FIGURE A.4 Air Sampling-Type Detector.

FIGURE A.5 Air-Purifying Respirator.

Air Sampling-Type Detector. A detector that consists of a piping or tubing distribution network that runs from the detector to the area(s) to be protected. (SEE FIGURE A.4.) *Note: An aspiration fan in the detector housing draws air from the protected area back to the detector through air sampling ports, piping, or tubing. At the detector, the air is analyzed for fire products.* NFPA 72, 2002 ed.

Air Tank. A storage vessel meeting the requirements of either U.S. Department of Transportation (DOT) or American Society of Mechanical Engineers (ASME) and used to store an accumulation of air under pressure. NFPA 1901, 2003 ed.

Air Traffic Control Provider. A service established to provide air and ground traffic control for airports. *Note: This includes airport control tower and airport flight information services.* NFPA 424, 2002 ed.

Air Transfer. The process of transferring air from one SCBA cylinder to another SCBA cylinder of the same rated pressure capacity by connecting them together with properly designed fittings and a high-pressure transfer line. NFPA 1500, 2002 ed.

Air Truck. A vehicle used to supply breathing air either to refill self-contained breathing apparatus (SCBA) or to supply respirators directly through hose lines. NFPA 1901, 2003 ed.

Air-Cushioned Vehicle (ACV). A vehicle that can travel on land and water. NFPA 402, 2002 ed.

Air-Mechanical Brakes. Brakes in which the force from an individual air chamber is applied directly to the friction surfaces through a mechanical linkage. NFPA 414, 2001 ed.

Air-Over-Hydraulic Brakes. Brakes in which the force of a master air cylinder is applied to the friction surfaces through an intervening hydraulic system. NFPA 414, 2001 ed.

Air-Purifying Respirator (APR). A respirator with an air-purifying filter, cartridge, or canister that removes specific air contaminants by passing ambient air through the air-purifying element. (SEE FIGURE A.5.) NFPA 1404, 2002 ed.

Airborne Emergency. Those emergencies that affect the operational integrity of an aircraft while in flight. *Note: The seriousness of these emergencies can be defined by using alert status guidelines stated in FAA terms, and aircraft emergencies for which services may be required, as defined in International Civil Aviation Organization Airport Services Manual, Part 1, "Rescue and Fire Fighting."* NFPA 424, 2002 ed.

Airborne Pathogens. Microorganisms that can produce infection and/or cause disease in humans after being inhaled. NFPA 1581, 2005 ed.

Aircraft Accident. An occurrence associated with the operation of an aircraft that takes place between the time any person boards the aircraft with the intention of flight and until all such persons have disembarked and in which any person suffers death or serious injury or in which the aircraft receives substantial damage. NFPA 403, 2003 ed.

Aircraft Accident Pre-Incident Planning. The process of forecasting all factors that could possibly exist involving an aircraft accident that could bear upon the existing emergency resources. *Note: A pre-incident plan should define the emergency organizational authority and the responsibilities of all those involved.* NFPA 402, 2002 ed.

Aircraft Emergency Exercise. Testing of the emergency plan and review of the results in order to improve the effectiveness of the plan. NFPA 424, 2002 ed.

Aircraft Familiarization. Refers to the knowledge of vital information that rescue and fire fighting personnel should learn and retain with regard to the specific types of aircraft that normally use the airport and other aircraft that might use the airport due to weather conditions at scheduled destinations. NFPA 402, 2002 ed.

Aircraft Fire Fighting. The control or extinguishment of fire adjacent to or involving an aircraft following ground accidents or incidents. *Note: Aircraft fire fighting does not include the control or extinguishment of airborne fires in aircraft.* NFPA 402, 2002 ed.

Aircraft Incident. An occurrence, other than an accident associated with the operation of an aircraft, that affects or could affect continued safe operation if not corrected. *Note: An incident does not result in serious injury to persons or substantial damage to aircraft.* NFPA 402, 2002 ed.

Aircraft Operator. A person, organization, or enterprise engaged in, or offering to engage in, aircraft operation. NFPA 424, 2002 ed.

Aircraft Rescue. The fire fighting action taken to prevent, control, or extinguish fire involving, or adjacent to, an aircraft for the purpose of providing maximum fuselage integrity and escape area for its occupants. *Note: Rescue and fire fighting personnel, to the extent possible, will assist in evacuation of the aircraft using normal and emergency means of egress. Additionally, rescue and fire fighting personnel will, by whatever means necessary, and to the extent possible, enter the aircraft and provide all possible assistance in the evacuation of the occupants.* NFPA 403, 2003 ed.

Aircraft Rescue and Fire Fighting Vehicle (ARFF). A vehicle intended to carry rescue and fire fighting equipment for rescuing occupants and combating fires in aircraft at, or in the vicinity of, an airport. NFPA 1002, 2003 ed.

Airline Coordinator. A representative authority delegated by an airline to represent its interests during an emergency. NFPA 424, 2002 ed.

Airport (Aerodrome). An area on land or water that is used or intended to be used for the landing and takeoff of aircraft and includes buildings and facilities. NFPA 402, 2002 ed.

Airport Air Traffic Control (ATC). A service established to provide air and ground traffic control for airports. NFPA 403, 2003 ed.

Airport Familiarization. Refers to the knowledge that rescue and fire fighting personnel must maintain relative to locations, routes, and conditions that will enable them to respond quickly and efficiently to emergencies on the airport and those areas surrounding the airport. NFPA 402, 2002 ed.

Airport Fire Chief. The individual normally having operational control over the airport's rescue and fire fighting personnel and equipment, or a designated appointee. NFPA 403, 2003 ed.

Airport Fire Department Personnel. Personnel under the operational jurisdiction of the chief of the airport fire department assigned to aircraft rescue and fire fighting. NFPA 403, 2003 ed.

Airport Fire Fighter. The Fire Fighter II who has demonstrated the skills and knowledge necessary to function as an integral member of an aircraft rescue and firefighting (ARFF) team. NFPA 1003, 2005 ed.

Airport Loading Walkway. An aboveground device through which passengers move between a point in an airport terminal building and an aircraft. *Note: Included in this category are walkways that are essentially fixed and permanently placed, or walkways that are essentially mobile in nature and that fold, telescope, or pivot from a fixed point at the airport terminal building.* NFPA 101, 2003 ed.

Airport Manager. The individual having managerial responsibility for the operation and safety of an airport. *Note: The manager can have administrative control over aircraft rescue and fire fighting services, but normally does not exercise authority over operational fire and rescue matters.* NFPA 403, 2003 ed.

Airport Terminal Building. A structure used primarily for air passenger enplaning or deplaning, including ticket sales, flight information, baggage handling, and other necessary functions in connection with air transport operation. *Note: This term includes any extensions and satellite buildings used for passenger handling or aircraft service functions. Aircraft loading walkways and "mobile lounges" are excluded.* NFPA 415, 2002 ed.

Airport/Community Emergency Plan. Establishment of procedures for coordinating the response of airport services with other agencies in the surrounding community that could be of assistance in responding to an emergency occurring on, or in the vicinity of, the airport. NFPA 424, 2002 ed.

Airside (Airport Operational Area). The movement area of an airport, adjacent terrain, and buildings or portions thereof, access to which is controlled. NFPA 424, 2002 ed.

Alarm. A functional mode in which the PASS alarm signal is activated by hand. NFPA 1982, 1998 ed.

Alarm Time. The point of receipt of the emergency alarm at the public safety answering point to the point where sufficient information is known to the dispatcher to deploy applicable units to the emergency. NFPA 1710, 2004 ed.

Alcohol-Resistant Foam. Used for fighting fires involving water-soluble materials or fuels that are destructive to other types of foams. *Note: Some alcohol-resistant foams may be capable of forming a vapor-suppressing aqueous film on the surface of hydrocarbon fuels.* NFPA 412, 2003 ed.

Alert Data Message (ADM). An analog or digital signal containing instructions for how a public alerting system alerting appliance (PASAA) is to deliver and, if capable, to acknowledge a public alert. NFPA 1221, 2002 ed.

All-Wheel Drive. A vehicle that drives on all wheels. NFPA 414, 2001 ed.

Alternate Air Source. A secondary air supply source system that involves an alternate second-stage regulator provided by either a separate dedicated second-stage or a multipurpose second-stage regulator coupled with a buoyancy compensator inflator valve. NFPA 1670, 2004 ed.

Alternative. A system, condition, arrangement, material, or equipment submitted to the authority having jurisdiction as a substitute for a requirement in a standard. NFPA 1144, 2002 ed.

Aluminum. A lightweight metal used extensively in the construction of aircraft airframes and skin sections. NFPA 402, 2002 ed.

Ambient. Someone's or something's surroundings, especially as they pertain to the local environment, for example, ambient air and ambient temperature. NFPA 921, 2004 ed.

Ambient Temperature. The temperature of the surrounding medium; usually used to refer to the temperature of the air in which a structure is situated or a device operates. NFPA 414, 2001 ed.

Ambulance. A vehicle designed, equipped, and operated for the treatment and transport of ill and injured persons. (See Figure A.6.) NFPA 450, 2004 ed.

Ambulance Service. An organization that exists to provide patient transportation by ambulance. NFPA 450, 2004 ed.

American College of Emergency Physicians (ACEP). A national organization of emergency medical physicians. NFPA 450, 2004 ed.

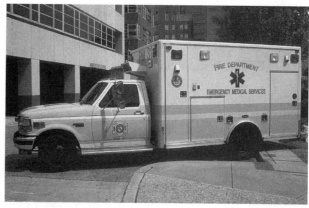

FIGURE A.6 Ambulance.

American Society for Nondestructive Testing (ASNT). A professional organization that is devoted to promoting knowledge of nondestructive testing. NFPA 1914, 2002 ed.

American Welding Society (AWS). An association that provides codes, guidelines, and standards utilized to evaluate welded structures and components in welded structures. NFPA 1914, 2002 ed.

Ampere. The unit of electric current that is equivalent to a flow of one coulomb per second; one coulomb is defined as 6.24×10^{18} electrons. NFPA 921, 2004 ed.

Anchor. A device designed to engage the bottom of a waterway and, through its resistance to drag, maintain a vessel within a given radius. NFPA 1925, 2004 ed.

Anchor Chocks. Fittings on a deck of a vessel used to stow an anchor when it is not in use. NFPA 1925, 2004 ed.

Anchor Point. A single, structural component used either alone or in combination with other components to create an anchor system capable of sustaining the actual and potential load on the rope rescue system. NFPA 1670, 2004 ed.

Anchor Rode. The line connecting an anchor with a vessel. NFPA 1925, 2004 ed.

Anchor Store. A department store or major merchandising center that has direct access to the covered mall but in which all required means of egress is independent of the covered mall. NFPA 101B, 2002 ed.

Anchor System. One or more anchor points rigged in such a way as to provide a structurally significant connection point for rope rescue system components. NFPA 1670, 2004 ed.

Anchorage. An area identified for safe anchoring. NFPA 1405, 2001 ed.

Ancillary Boom Ladder. A ladder or ladders affixed to a telescoping or articulating boom section. NFPA 1914, 2002 ed.

Angle of Approach. The smallest angle made between the road surface and a line drawn from the front point of ground contact of the front tire to any projection of the apparatus in front of the front axle. NFPA 1901, 2003 ed.

Angle of Departure. The smallest angle made between the road surface and a line drawn from the rear point of ground contact of the rear tire to any projection of the apparatus behind the rear axle. NFPA 1901, 2003 ed.

Angle of Inclination. The angle incorporated between the beams and a level plane. NFPA 1931, 2004 ed.

Angle of Repose. The greatest angle above the horizontal plane at which loose material (such as soil) will lie without sliding. NFPA 1670, 1999 ed.

Annunciator. The component of a PASS device designed to emit audible signals. NFPA 1982, 1998 ed.

ANSI/ASME. The designation for American National Standards Institute publication, sponsored and published by the American Society of Mechanical Engineers. NFPA 1, 2003 ed.

Antenna. A device connected to a receiver, transmitter, or transceiver that radiates the transmitted signal, receives a signal, or both. NFPA 1221, 2002 ed.

Apartment Building. A building or portion thereof containing three or more dwelling units with independent cooking and bathroom facilities. NFPA 5000, 2002 ed.

Apparatus. *As applied to fire fighting:* A motor-driven vehicle or group of vehicles designed and constructed for the purpose of fighting fires. NFPA 1143, 2003 ed. *As applied to laboratories:* Furniture, chemical fume hoods, centrifuges, refrigerators, and commercial or made-on-site equipment used in a laboratory. NFPA 45, 2004 ed.

Applicable Codes and Standards. Those codes and standards that are legally adopted and enforced by a jurisdiction at the time of construction of an occupancy or installation of a system or of equipment. *Note: These applicable codes and standards can include ordinances, statutes, regulations, or other legal documents adopted by the jurisdiction.* NFPA 1031, 2003 ed.

Approach Assessment. The period of time from the moment when the incident site first becomes visible to the moment when the initial size-up is completed. NFPA 1670, 1999 ed.

Approach Fire Fighting. Limited, specialized exterior fire fighting operations at incidents involving fires producing very high levels of conductive, convective, and radiant heat, such as bulk flammable gas and bulk flammable liquid fires. NFPA 1500, 2002 ed.

Arc. A high-temperature luminous electric discharge across a gap or through a medium such as charred insulation. NFPA 921, 2004 ed.

Arch. The bottom curve of the foot from the heel to the ball. NFPA 1951, 2005 ed.

Arcing Through Char. Arcing associated with a matrix of charred material (e.g., charred conductor insulation) that acts as a semiconductive medium. NFPA 921, 2004 ed.

Area of Origin. The room or area where a fire began. NFPA 921, 2004 ed.

ARFF Chassis. The assembled frame, engine, drive train, and tires of a vehicle. NFPA 414, 2001 ed.

ARFF Personnel. Personnel under the operational jurisdiction of the chief of the airport fire department assigned to aircraft rescue and fire fighting. NFPA 405, 2004 ed.

Arrival. The point at which a vehicle is stopped on the scene of a response destination or address. NFPA 450, 2004 ed.

Arrow Pattern. A fire pattern displayed on the cross section of a burned wooden structural member. NFPA 921, 2004 ed.

Arson. The crime of maliciously and intentionally, or recklessly, starting a fire or causing an explosion. NFPA 921, 2004 ed.

Articulating Boom. An aerial device consisting of two or more folding boom sections whose extension and retraction modes are accomplished by adjusting the angle of knuckle joints. NFPA 1901, 2003 ed.

Artificial Barricade. An artificial mound or revetted wall of earth of a minimum thickness of 0.9 m (3 ft). NFPA 1125, 2001 ed.

Artificial Sea Water. A solution that consists of 1.10 percent magnesium chloride hexahydrate, 0.16 percent calcium chloride dihydrate, 0.40 percent anhydrous sodium sulfate, 2.50 percent sodium chloride, and 95.84 percent deionized or distilled water. NFPA 1150, 2004 ed.

Ascending (Line). A means of safely traveling up a fixed line with the use of one or more ascent devices. NFPA 1670, 2004 ed.

Ascent Device. An auxiliary equipment system component; a friction or mechanical device utilized to allow ascending a fixed line. NFPA 1670, 2004 ed.

ASME. American Society of Mechanical Engineers. NFPA 55, 2005 ed.

ASME Pressure Vessel. A pressure vessel used for the storage or accumulation of air or gas under pressure that is constructed and tested in accordance with the ASME Boiler and Pressure Vessel Code. NFPA 1901, 2003 ed.

Aspect. Compass direction toward which a slope faces. NFPA 1144, 2002 ed.

Aspirate. To draw in air; nozzle-aspirating systems draw air into the nozzle to mix with the foam solution. NFPA 1145, 2000 ed.

Assembly Occupancy. An occupancy (1) used for a gathering of 50 or more persons for deliberation, worship, entertainment, eating, drinking, amusement, awaiting transportation, or similar uses; or (2) used as a special amusement building, regardless of occupant load. NFPA 101, 2003 ed.

Assessment. A structured process by which relevant information is gathered for the purpose of determining specific child or family intervention needs. NFPA 1035, 2005 ed.

Assessment Phase (Size-Up). The process of assessing the conditions, the scene, and the subject's condition and ability to assist in his or her own rescue. NFPA 1670, 2004 ed.

Assistant Safety Officer. A member of the fire department assigned and authorized by the fire chief to assist the fire department health and safety officer or incident safety officer in the performance of duties and responsibilities. NFPA 1521, 2002 ed.

Athwartship. Side to side, at right angles to fore and aft centerline of a ship. NFPA 1405, 2001 ed.

Atmosphere-Supplying Respirator. A respirator that supplies the respirator user with breathing air from a source independent of the ambient atmosphere, which includes self-contained breathing apparatus (SCBA) and supplied air respirators (SAR). NFPA 1852, 2002 ed.

Atmospheric Monitoring. A method of evaluating the ambient atmosphere of a space, including but not limited to its oxygen content, flammability, and toxicity. NFPA 1006, 2003 ed.

Atmospheric Pressure. The pressure of the weight of air and water vapor on the surface of the earth, approximately 14.7 pounds per square inch (psia) (101 kPa absolute) at sea level. NFPA 54, 2002 ed.

Attached Building. A building having only one common wall with another building having other types of occupancies. NFPA 30, 2003 ed.

Attack Hose. Hose designed to be used by trained fire fighters and fire brigade members to combat fires beyond the incipient stage. NFPA 1961, 2002 ed.

Attack Line. A hose line used primarily to apply water directly onto a fire and operated by a sufficient number of personnel so that it can be maneuvered effectively and safely. NFPA 1410, 2005 ed.

Attack Pump. A centrifugal water pump permanently mounted on the apparatus with a rated capacity of 250 gpm (950 L/min) or more, but less than 750 gpm (2850 L/min), at 150 psi (1035 kPa) net pump pressure, and used for fire fighting. NFPA 1002, 1998 ed.

Attendant. A term used to describe U.S. federally regulated industrial workers who are qualified to be stationed outside one or more confined spaces, who monitor authorized entrants, and who perform all of the following duties: (1) remain outside the confined space during entry operations until relieved by another attendant, (2) summon rescue and other needed resources as soon as the attendant determines that authorized entrants might need assistance to escape from confined space hazards, (3) perform nonentry rescues as specified by the rescue procedure listed on the permit. NFPA 1670, 1999 ed.

Attic Extension Ladder. An extension ladder that is specifically designed to be used to gain entry through a scuttle, hatch, or other similarly restricted opening. NFPA 1931, 2004 ed.

Authority Having Jurisdiction. An organization, office, or individual responsible for enforcing the requirements of a code or standard, or for approving equipment, materials, an installation, or a procedure. NFPA Official Definition.

Authorized Entrant. Describes U.S. federally regulated industrial workers who are designated to enter confined spaces and who meet the following training requirements for each specific space they enter: (a) Hazard Recognition. The ability to recognize the signs and symptoms of exposure to a hazardous material or atmosphere within the space and to understand the consequences of exposure and the mode of transmission (i.e., injection, ingestion, inhalation, or absorption) for the hazard. (b) Communications. The ability to carry out the method by which rescue services are to be summoned in the event of an emergency, the method by which the entrant will communicate with the attendant on the outside of the space, and a backup method of communication should the primary system fail. (c) Personal Protective Equipment (PPE). The ability to use all PPE appropriate for the confined space. (d) Self-Rescue. The ability to carry out the method by which the entrant will escape from the space should an emergency occur. NFPA 1670, 1999 ed.

Authorized Person. A person approved or assigned to perform specific types of duties or to be at a specific location at the job site. NFPA 1901, 2003 ed.

Autoignition. Initiation of combustion by heat but without a spark or flame. NFPA 921, 2004 ed.

Autoignition Temperature. The lowest temperature at which a combustible material ignites in air without a spark or flame. NFPA 921, 2004 ed.

Automated External Defibrillator (AED). A device that administers an electric shock through the chest wall to the heart using built-in computers to assess the patient's heart rhythm and defibrillate as needed. NFPA 450, 2004 ed.

Automated Vehicle Locator (AVL). A computerized mapping system used to track the location of vehicles. NFPA 450, 2004 ed.

Automatic Aid. A plan developed between two or more fire departments for immediate joint response on first alarms. NFPA 1142, 2001 ed.

Automatic Electrical Load Management System. A device that continuously monitors the electrical system voltage and sheds predetermined loads in a selected order to prevent overdischarging of the apparatus' batteries. NFPA 1911, 2002 ed.

Automatic Fire Extinguishing System. Any system that is designed and installed to detect a fire and subsequently discharge an extinguishing agent without human activation or direction. NFPA 1141, 2003 ed.

Automatic Locking Differential. A type of nonslip differential that operates automatically. NFPA 414, 2001 ed.

Automatic Regulating Proportioning System. A proportioning system that automatically adjusts the flow of foam concentrate into the water stream to maintain the desired proportioning ratio. NFPA 1906, 2001 ed.

Automotive Service Stations. That portion of a property where liquids used as motor fuels are stored and dispensed from fixed equipment into the fuel tanks of motor vehicles or approved containers and shall include any facilities for the sale and service of tires, batteries, and accessories. *Note: This occupancy designation also applies to buildings, or portions of buildings, used for lubrication, inspection, and minor automotive maintenance work, such as tune-ups and brake system repairs. Major automotive repairs, painting, and body and fender work are excluded.* NFPA 1, 2000 ed.

Auxiliary Braking System. A braking system in addition to the service brakes, such as an engine retarder, transmission retarder, driveline retarder, or exhaust retarders. NFPA 1901, 2003 ed.

Auxiliary Engine-Driven Pumps. Pumps whose power is provided by engines that are independent of the vehicle engine. NFPA 1906, 2001 ed.

Auxiliary Equipment. System components that are load-bearing accessories designed to be utilized with life safety rope and harness including, but not limited to, ascending devices, carabiners, descent control devices, rope grab devices, and snap-links. NFPA 1670, 2004 ed.

Auxiliary Hydraulic Power. A small gasoline engine, diesel engine, or electric motor-driven hydraulic pump used to operate an aerial device in an emergency or in lieu of the main hydraulic system. NFPA 1901, 2003 ed.

Auxiliary Power Unit (APU). A self-contained power source, provided as a component of an aircraft, that is used to energize aircraft systems when power plants are not operating or when external power is not available. NFPA 402, 2002 ed.

Auxiliary Pump. A water pump mounted on the fire apparatus in addition to a fire pump and used for fire fighting either in conjunction with or independent of the fire pump. NFPA 1901, 2003 ed.

Avalanche. A mass of snow—sometimes containing ice, water, and debris—that slides down a mountainside. NFPA 1670, 2004 ed.

Axle Tread. The distance between the center of two tires or wheels on one axle. NFPA 414, 2001 ed.

B-Class Division. A fire barrier system consisting of bulkheads or decks and including all penetrations for piping and cables, doors, windows, and ductwork, providing 30 minutes of fire resistance when tested in accordance with established test methods. NFPA 301, 2001 ed.

Back Length. Upper torso garment measurement at center back from bottom of collar to bottom edge of garment. NFPA 1977, 2005 ed.

Back Rise. Lower torso garment measurement from crotch seam to top of waistband at back center. NFPA 1977, 2005 ed.

Back Stock Area. The area of a mercantile occupancy that is physically separated from the sales area and not intended to be accessible to the public. NFPA 30B, 2002 ed.

Back-Up Alarm. An audible device designed to warn that the fire apparatus is in reverse gear. NFPA 1901, 2003 ed.

Backdraft. A phenomenon that occurs when a fire takes place in a confined area such as a sealed aircraft fuselage and burns undetected until most of the oxygen within is consumed. The heat continues to produce flammable gases, mostly in the form of carbon monoxide. These gases are heated above their ignition temperature. When a supply of oxygen is introduced, as when normal entry points are opened, the gases could ignite with explosive force. (See Figure B.1.) NFPA 402, 2002 ed.

Backfire. A fire set along the inner edge of a fire control line to consume the fuel in the path of a wildland fire or change the direction of force of the fire's convection column. NFPA 901, 2001 ed.

Backup Line. An additional hose line used to reinforce and protect personnel in the event the initial attack proves inadequate. NFPA 1410, 2005 ed.

Ballast. Weight, liquid or solid, added to a ship to ensure stability. NFPA 1405, 2001 ed.

Ballast Tank. A watertight compartment to hold liquid ballast. NFPA 1405, 2001 ed.

Band. A range of frequencies between two definite limits. NFPA 1221, 2002 ed.

Barge. A long, large vessel, usually flat-bottomed, self-propelled, towed, or pushed by another vessel, used for transporting materials. NFPA 1405, 2001 ed.

Bark Pocket Wood Irregularity. An opening between annual growth rings that contains bark. NFPA 1931, 2004 ed.

Barricade. A natural or artificial barrier that effectively screens a magazine, building, railway, or highway from the effects of an explosion in a magazine or building containing explosives. NFPA 1124, 2003 ed.

Barrier Layer. The layer of garment material, glove material, footwear material, or face protection device material designated as providing blood and body fluid-borne pathogen resistance. NFPA 1999, 2003 ed.

Barrier Material. A single-layer fabric or a laminated or coated, multilayer material considered as a single-layer fabric that limits transfer from the face of the layer to the other side. NFPA 1971, 2000 ed.

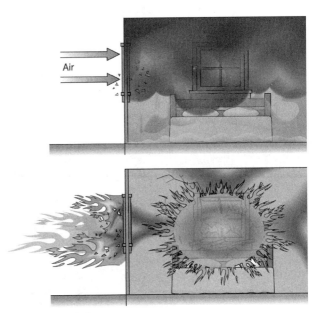

Figure B.1 Backdraft.

Base (Bed) Section. The lowest or widest section of an extension ladder. NFPA 1931, 2004 ed.

Base Rail. The lower chord (rail) of an aerial ladder to which rungs and reinforcements are attached. NFPA 1901, 2003 ed.

Base Section. The first or bottom section of an aerial device. NFPA 1901, 2003 ed.

Base Station. A stationary radio transceiver with an integral AC power supply. (See Figure B.2.) NFPA 1221, 2002 ed.

Basement. Story of a building wholly below grade or partly below and partly above grade, located so that the vertical distance from grade to the floor below is greater than the vertical distance from grade to the floor above. NFPA 1, 2003 ed.

Basement and Underground Parking Structures. Parking structures that are located below grade. *Note: A basement parking structure has other occupancies above it; an underground parking structure has no occupancy other than parking above it. Basement and underground parking structures are considered as specific cases of enclosed parking structures.* NFPA 1, 2003 ed.

Basic First Aid Kit. Equipment or devices for managing infection exposure, airways, spinal immobilization, fracture immobilization, shock, and bleeding control. NFPA 1006, 2003 ed.

Basic Life Support (BLS). Emergency medical treatment at a level as defined by the medical authority having jurisdiction. NFPA 1500, 2002 ed.

Basic Plane. The anatomical plane that includes the superior rim of the external auditory meatus, the upper edge of the external opening of the ear, and the inferior margin of the orbit, which is the lowest point of the floor of the eye socket. NFPA 1971, 2000 ed.

Basic Spray Nozzle. An adjustable-pattern spray nozzle in which the rated discharge is delivered at a designated nozzle pressure and nozzle setting. NFPA 1964, 2003 ed.

Basic Weight. The weight of the helmet, including the following components: shell; energy-absorbing system; retention system; fluorescent and retroflective trim; ear covers; and a faceshield, goggles, or both. NFPA 1971, 2000 ed.

Batch Mix. The manual addition of foam concentrate to a water storage container or tank to make foam solution. NFPA 1145, 2000 ed.

Battery System (Lead-Acid). A system that consists of these interconnected subsystems: (1) Lead-acid batteries; (2) Battery chargers; (3) A collection of rectifiers, inverters, converters, and associated electrical equipment as required for a particular application. NFPA 1, 2003 ed.

Battery, Valve-Regulated (VRLA). A lead-acid battery consisting of sealed cells furnished with a valve that opens to vent the battery whenever the internal pressure of the battery exceeds the ambient pressure by a set amount. NFPA 1, 2003 ed.

Battery, Vented (Flooded). A lead-acid battery consisting of cells that have electrodes immersed in liquid electrolyte. NFPA 1, 2003 ed.

Baud. A unit of signaling speed equal to the number of discrete conditions or signal events per second. NFPA 1221, 2002 ed.

Bead. A rounded globule of re-solidified metal at the end of the remains of an electrical conductor that was caused by arcing and is characterized by a sharp line of demarcation between the melted and unmelted conductor surfaces. NFPA 921, 2004 ed.

Beam. The breadth (i.e., width) of a ship at its widest point. NFPA 1405, 2001 ed.

Beam (Side Rail). The main structural side of the ground ladder. NFPA 1931, 2004 ed.

Belay. The method by which a potential fall distance is controlled to minimize damage to equipment and/or injury to a live load. NFPA 1670, 2004 ed.

Belayer. The rescuer who operates the belay system. NFPA 1006, 2003 ed.

Bell-Bottom Pier Hole. A type of shaft or footing excavation, the bottom of which is made larger than the cross section above to form a bell shape. NFPA 1670, 2004 ed.

Belt. A system component; material configured as a device that fastens around the waist only and designated as a ladder belt, an escape belt, or a ladder/escape belt. (See Figure B.3.) NFPA 1006, 2003 ed.

FIGURE B.2 Base Station.

FIGURE B.3 Ladder Belt.

Benching or Benching System. A method of protecting employees from cave-ins by excavating the side of an excavation to form one or a series of horizontal levels or steps, usually with vertical or near-vertical surfaces between levels. NFPA 1670, 2004 ed.

Bend. A knot that joins two ropes or webbing pieces together. (See Figure B.4.) NFPA 1670, 2004 ed.

Beneficial System. Auxiliary-powered equipment in motor vehicles or machines that can enhance or facilitate rescues such as electric, pneumatic, or hydraulic seat positioners, door locks, window operating mechanisms, suspension systems, tilt steering wheels, convertible tops, or other devices or systems to facilitate the movement (extension, retraction, raising, lowering, conveyor control) of equipment or machinery. NFPA 1006, 2003 ed.

Berth. (1) The mooring of a boat alongside a bulkhead, pier, or between piles. (2) A sleeping space. NFPA 1405, 2001 ed.

Berthing Area. (1) A bed or bunk space on a ship. (2) A space at a wharf for docking a ship. NFPA 1405, 2001 ed.

Bight. The open loop in a rope or piece of webbing formed when it is doubled back on itself. (See Figure B.5.) NFPA 1670, 1999 ed.

Bilge. The lowest inner part of a ship's hull. NFPA 1405, 2001 ed.

Biodegradability. A measure of the decomposition of organic matter through the action of microorganisms. NFPA 1150, 2004 ed.

Biological Agents. Biological materials that are capable of causing an acute disease or long-term damage to the human body. NFPA 1951, 2005 ed.

Biological Warfare Agent. A biological substance intended to kill, seriously injure, or incapacitate humans through physiological effects. NFPA 1991, 2005 ed.

Bit. The smallest unit of computer storage. NFPA 1221, 2002 ed.

Bitragion Coronal Arc. The arc between the right and left tragion as measured over the top of the head in a plane perpendicular to the midsagittal plane. (See Figure B.6.) NFPA 1971, 2000 ed.

Bitragion Inion Arc. The arc between tragion as measured over the inion. (See Figure B.7.) NFPA 1971, 2000 ed.

Bitt. Any of the deck posts, often found in pairs, around which ropes or cables are wound and held fast. NFPA 1925, 2004 ed.

FIGURE B.4 Bend.

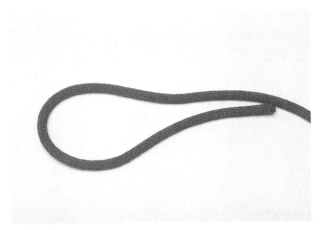

FIGURE B.5 Bight.

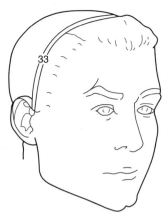

FIGURE B.6 Bitragion Coronal Arc.

Bitter End. That end of a rope or cable that is wound around a bitt, for example, the onboard end of the anchor rode, which is usually permanently attached to the vessel. NFPA 1925, 2004 ed.

Bladder Fuel Tank. A fuel container that is both collapsible and self-sealing. NFPA 410, 2004 ed.

Blanking and Blinding. A form of hydraulic energy isolation that is the absolute closure of a pipe, line, or duct by fastening a solid plate (such as a spectacle blind or skillet blind) that completely covers the bore and that is capable of withstanding the maximum pressure within the pipe, line, or duct with no leakage beyond the plate. NFPA 1670, 1999 ed.

Blast Pressure Front. The expanding leading edge of an explosion reaction that separates a major difference in pressure between normal ambient pressure ahead of the front and potentially damaging high pressure at and behind the front. NFPA 921, 2004 ed.

BLEVE. Boiling liquid expanding vapor explosion. (See Figure B.8.) NFPA 921, 2004 ed.

Block Creel Construction. Rope constructed without knots or splices in the yarns, ply yarns, strands or braids, or rope. NFPA 1983, 2001 ed.

Blood. Human blood, human blood components, and products made from human blood. NFPA 1581, 2005 ed.

Blood and Body Fluid-Borne Pathogen. An infectious bacteria or virus carried in human, animal, or clinical body fluids, organs, or tissues. NFPA 1999, 2003 ed.

Bloodborne Pathogens. Pathogenic microorganisms that are present in human blood and can cause diseases in humans. NFPA 450, 2004 ed.

Blunt Start. The removal of the incomplete thread at the end of the thread. *Note: This is a feature of threaded parts that are repeatedly assembled by hand. Also known as the "Higbee cut."* NFPA 1963, 1998 ed.

Board of Appeals. A group of persons appointed by the governing body of the jurisdiction adopting a code for the purpose of hearing and adjudicating differences of opinion between the authority having jurisdiction and the citizenry in the interpretation, application, and enforcement of the Code. NFPA 1, 2003 ed.

Boarding Ladder. A device used for boarding a vessel from the water, including handles, rails, ladders, steps, or platforms. NFPA 1925, 2004 ed.

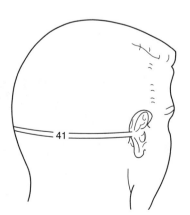

FIGURE B.7 Bitragion Inion Arc.

FIGURE B.8 BLEVE.

Body Fluids. Fluids that the body produces including, but not limited to, blood, semen, mucus, feces, urine, vaginal secretions, breast milk, amniotic fluids, cerebrospinal fluid, synovial fluid, pericardial fluid, sputum, and any other fluids that might contain pathogens. NFPA 1581, 2005 ed.

Body Recovery. An operation involving the retrieval of the remains of a deceased victim, but in no case a living person. NFPA 1670, 2004 ed.

Body Waste. The discharge from any fixture, appliance, or appurtenance containing fecal matter or urine. NFPA 1192, 2005 ed.

Bogie. *As applied to aircraft:* A tandem arrangement of aircraft landing gear wheels. *Note: The bogie can swivel up and down so that all wheels follow the ground as the attitude of the aircraft changes or the ground surface changes.* NFPA 402, 2002 ed. *As applied to fire fighting and rescue vehicles:* A combination of two axles used to support the end of a vehicle. NFPA 414, 2001 ed.

Boiling Point. The temperature at which the vapor pressure of a liquid equals the surrounding atmospheric pressure. *Note: For purposes of defining the boiling point, atmospheric pressure is considered to be 14.7 psia (760 mm Hg). For mixtures that do not have a constant boiling point, the 20 percent evaporated point of a distillation performed in accordance with ASTM D 86, Standard Test Method for Distillation of Petroleum Products at Atmospheric Pressure, is considered to be the boiling point.* NFPA 30, 2003 ed.

Bombproof. Refers to a single anchor point capable of sustaining the actual or potential forces exerted on the rope rescue system without possibility of failure. NFPA 1006, 2003 ed.

Bonded Warehouse. A facility for storing controlled materials, such as alcohol and tobacco products, that generally has high levels of security and very limited access. NFPA 1620, 2003 ed.

Boom. *As applied to fire fighting apparatus:* An assembled section of an aerial device. *Note: The boom construction can be of the stressed skin box beam type, the trussed-lattice-type, or the open 'U' truss-type design.* NFPA 1901, 2003 ed. *As applied to marine vessels:* (1) A long pole extending upward at an angle from the mast of a derrick to support or guide objects lifted or suspended. (2) A floating barrier used to confine materials upon the surface of the water (e.g., oil). NFPA 1405, 2001 ed.

Boom Boost Cylinders. The hydraulic cylinders located on the upper boom of an articulating boom aerial device that help lift the upper boom from the lower boom. NFPA 1914, 2002 ed.

Boom Support. A structural component that is attached to the chassis frame and that is used to support the aerial device when it is in the cradled position. NFPA 1914, 2002 ed.

Booster Hose. A non-collapsible hose used under positive pressure having an elastomeric or thermoplastic tube, a braided or spiraled reinforcement, and an outer protective cover. NFPA 1962, 2003 ed.

Booster Supplied Air System. A system that is capable of increasing air pressure from an air storage system or a compressor system. NFPA 1901, 2003 ed.

Bottom Circumference. Measurement of upper or lower torso garment along bottom edge of the garment from folded edge to folded edge, and multiplied by 2 to obtain circumference. NFPA 1977, 2005 ed.

Bottom Shield. A protective layer installed between the floor and Type FCC flat conductor cable to protect the cable from physical damage and may or may not be incorporated as an integral part of the cable. NFPA 70, 2005 ed.

Bow. *As applied to fire department aerial devices:* The distance that the end of an aerial ladder or boom deviates from a straight line extension of the base section. NFPA 1914, 2002 ed. *As applied to marine vessels:* The front end of boat or vessel. NFPA 1405, 2001 ed.

Box Circuit. A circuit that is connected to boxes that transmit an alarm to the communications center. NFPA 1221, 2002 ed.

Braided Reinforcement. A hose reinforcement consisting of one or more layers of interlaced spiraled strands of yarn or wire, with a layer of rubber between each braid. NFPA 1962, 2003 ed.

Branch. A supervisory level established in either the operations or logistics function to provide a span of control. NFPA 1561, 2005 ed.

Breach. An opening made in the wall, floor, or ceiling of a structure, based on construction type, that can be used for moving rescuers, equipment, or victims into or out of the structure. NFPA 1006, 2003 ed.

Breaching Techniques. Methods that utilize breaking and cutting tools to create safe openings in masonry, concrete, and wood structures. NFPA 1006, 2003 ed.

Break Bulk Terminal. A terminal where commodities packaged in bags, drums, cartons, and crates are commonly, but not always, palletized and loaded and unloaded. NFPA 1405, 2001 ed.

Break-Apart Monitor. A monitor that can be converted for use either in stationary mode on a fire apparatus or in portable mode on a separate ground base. NFPA 1965, 2003 ed.

Breakover. A fire edge that crosses a control line or natural barrier intended to confine the fire and the resultant fire. NFPA 1051, 2002 ed.

Breathing Air. Air that meets the requirements specified in CGA G-7.1, Commodity Specification for Air, for Grade D and E air for human respiration and NFPA 1500, Standard on Fire Department Occupational Safety and Health Program. NFPA 1989, 2003 ed.

Breathing Air Cylinder. The pressure vessel or vessels that are an integral part of the SCBA and that contain the breathing gas supply; can be configured as a single cylinder or other pressure vessel, or as multiple cylinders or pressure vessels. NFPA 1981, 2002 ed.

Breathing Air System. The complete assembly of equipment such as compressors, purification system, pressure regulators, safety devices, manifolds, air tanks or receivers, and interconnected piping required to deliver air for breathing. NFPA 1901, 2003 ed.

Bridge. The vessel's command and control area, usually enclosed, containing the principal helm, navigation systems, communications systems, and monitoring equipment for the vessel's operating systems. *Note: Also called the pilothouse.* NFPA 1925, 2004 ed.

Brim. The part of the helmet shell extending around the entire circumference of the helmet. NFPA 1971, 2000 ed.

Brim Line. The horizontal plane intersecting the point of the front opening of the helmet at the midsagittal plane. NFPA 1971, 2000 ed.

British Thermal Unit (Btu). The quantity of heat required to raise the temperature of one pound of water 1°F at the pressure of 1 atmosphere and temperature of 60°F. *Note: A British thermal unit is equal to 1055 joules, 1.055 kilojoules, and 252.15 calories.* NFPA 921, 2004 ed.

Brush. A collective term that refers to a stand of vegetation dominated by shrubs, woody plants, or low-growing trees, usually of a type undesirable for livestock or timber management. NFPA 1143, 2003 ed.

Buckle. A load-bearing connector that is an integral part of an auxiliary equipment system component and used to connect two pieces of webbing. NFPA 1983, 2001 ed.

Building. Any structure used or intended for supporting or sheltering any use or occupancy. NFPA 101B, 2002 ed.

Building Construction. Types of construction based on the combustibility and the fire resistance rating of a building's structural elements. NFPA 1051, 2002 ed.

Building Service Equipment. The items or components that provide lighting, heating, ventilation, and air conditioning along with elevators and escalators. NFPA 1031, 2003 ed.

Bulk Merchandising Retail Building. A building in which the sales area includes the storage of combustible materials on pallets, in solid piles, or in racks in excess of 3.7 m (12 ft) in storage height. NFPA 5000, 2002 ed.

Bulk Packaging. Any packaging, including transport vehicles, having a liquid capacity of more than 450 L (119 gal), a solids capacity of more than 400 kg (882 lb), or a compressed gas water capacity of more than 454 kg (1001 lb). NFPA 472, 2002 ed.

Bulk Plant or Terminal. That portion of a property where liquids are received by tank vessel, pipeline, tank car, or tank vehicle and are stored or blended in bulk for the purpose of distributing such liquids by tank vessel, pipeline, tank car, tank vehicle, portable tank, or container. NFPA 30, 2003 ed.

Bulk Solid Storage. The storage of more than 2722 kg (6000 lb) in a single container. NFPA 430, 2004 ed.

Bulk Terminal. A terminal where unpackaged commodities carried in the holds and tanks of cargo vessels and tankers and generally transferred by such means as conveyors, clamshells, and pipelines are handled. NFPA 1405, 2001 ed.

Buoyancy. (1) The tendency or capacity to remain afloat in a liquid. (2) The upward force of a fluid upon a floating object. NFPA 1405, 2001 ed.

Buoyancy Control Device. Jacket or vest that contains an inflatable bladder for the purposes of controlling buoyancy. NFPA 1006, 2003 ed.

Burning. The process of self-perpetuating combustion, with or without an open flame. *Note: Smoldering is burning.* NFPA 901, 2001 ed.

Burning Out. Setting fire inside a control line to consume the fuel between the edge of the fire and the control line. NFPA 1051, 2002 ed.

Burning Velocity. The rate of flame propagation relative to the velocity of the unburned gas that is ahead of it. NFPA 68, 2002 ed.

Burst Pressure. The pressure at which a hydraulic component fails due to stresses induced as a result of the pressure. NFPA 1901, 2003 ed.

Burst Test Pressure. A pressure equal to at least three times the service test pressure. NFPA 1961, 2002 ed.

Business Continuity Program. An ongoing process supported by senior management and funded to ensure that the necessary steps are taken to identify the impact of potential losses, maintain viable recovery strategies and recovery plans, and ensure continuity of services through personnel training, plan testing, and maintenance. NFPA 1600, 2004 ed.

Business Occupancy. An occupancy used for account and recordkeeping or the transaction of business other than mercantile. NFPA 5000, 2002 ed.

Butt. The end of the beam placed on the ground, or other lower support surface, when ground ladders are in the raised position. NFPA 1931, 2004 ed.

Butt Spurs (Feet). That component of ground ladder support that is in contact with the lower support surface to reduce slippage. *Note: Butt spurs can be the lower end of beams, or can be added devices.* NFPA 1931, 2004 ed.

Byte. The common unit of computer storage. NFPA 1221, 2002 ed.

Cable. A wire rope used to transmit forces from one component to another for the purpose of extending or retracting an aerial device. NFPA 1901, 1999 ed.

Cable. A factory assembly of two or more conductors having an overall covering. NFPA 70, 2005 ed.

Cable Assembly. A powered rescue tool component consisting of the power cable with all permanently attached connectors that connect the powered rescue tool to the power unit. NFPA 1936, 2005 ed.

Cable Separation Guide. The mechanism that aligns and separates the cable when it is wound on the drum of an aerial ladder's extension winch. NFPA 1914, 2002 ed.

Calibrate. To correlate the reading of an instrument or system of measurement with a standard. NFPA 1915, 2000 ed.

Call Detail Recording (CDR). A system that provides a record of each call, including automatic number identification (ANI), trunk number, and answering attendant number; and the time of seizure, answer, and disconnect/transfer. NFPA 1221, 2002 ed.

Calorie. The amount of heat necessary to raise 1 gram of water 1°C at 15°C. *Note: A calorie is 4.184 joules, and there are 252.15 calories in a British thermal unit (Btu).* NFPA 921, 2004 ed.

Campaign. A component of an organizational fire and life safety education strategy with a predetermined time frame. NFPA 1035, 2005 ed.

Candidate. *As applied to fire inspector or plan examiner:* A person who has applied to become a fire inspector or plan examiner. NFPA 1031, 2003 ed. *As applied to fire departments:* A person who has submitted an application to become a member of the fire department. NFPA 1500, 2002 ed.

Captive. A firm or group that forms an insurance company for its own purposes. NFPA 1250, 2004 ed.

Car Terminal. A terminal where automobiles are the commodity handled. NFPA 1405, 2001 ed.

Carabiner. An auxiliary equipment system component; an oval or D-shaped metal, load-bearing connector with a self-closing gate used to join other components of a rope system. (See Figure C.1.) NFPA 1983, 2001 ed.

Carbon Monoxide Monitor. A monitoring device that samples a purified air stream for trace elements of carbon monoxide (CO). (See Figure C.2.) NFPA 1901, 2003 ed.

Carcinogen/Carcinogenic. A cancer-causing substance that is identified in one of several published lists. NFPA 1851, 2001 ed.

Care. Procedures for cleaning, decontamination, and storage of protective clothing and equipment. NFPA 1851, 2001 ed.

Cargo Pockets. Pockets located on the protective garment exterior. NFPA 1976, 2000 ed.

Cascade System. A method of piping air tanks together to allow air to be supplied to the SCBA fill station using a progressive selection of tanks, each with a higher pressure level. NFPA 1901, 2003 ed.

Category A Medical Condition. A medical condition that would preclude a person from performing as a member in a training or emergency operational environment by presenting a significant risk to the safety and health of the person or others. NFPA 1582, 2003 ed.

Figure C.1 Carabiner.

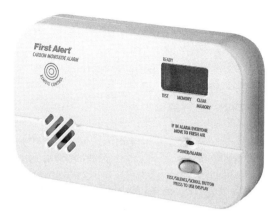

Figure C.2 Carbon Monoxide Monitor.

Category B Medical Condition. A medical condition that, based on its severity or degree, could preclude a person from performing as a member in a training or emergency operational environment by presenting a significant risk to the safety and health of the person or others. NFPA 1582, 2003 ed.

Cathode. A metal that in an electrolyte assumes a more electropositive charge than one to which it is coupled. *Note: This metal tends not to corrode or dissolve in an electrolyte.* NFPA 1925, 2004 ed.

Cause. The circumstances, conditions, or agencies that brought about or resulted in the fire or explosion incident, damage to property resulting from the fire or explosion incident, or bodily injury or loss of life resulting from the fire or explosion incident. NFPA 921, 2004 ed.

Cave-In. The separation of a mass of soil or rock material from the side of an excavation or trench, or the loss of soil from under a trench shield or support system, and its sudden movement into the excavation, either by falling or sliding, in sufficient quantity so that it could entrap, bury, or otherwise injure and immobilize a person. NFPA 1670, 2004 ed.

Ceiling Layer. A buoyant layer of hot gases and smoke produced by a fire in a compartment. (See Figure C.3.) NFPA 921, 2004 ed.

Census Data. Demographic population data available by statistical areas from a governmental agency. NFPA 901, 2001 ed.

Center of Gravity. The point at which the entire weight of the fire apparatus is considered to be concentrated so that, if supported at this point, the apparatus would remain in equilibrium in any position. NFPA 1901, 2003 ed.

Centerline. A line that runs from the bow to the stern of the vessel and is equidistant from the port and starboard sides of the vessel. NFPA 1405, 2001 ed.

Central Station Fire Alarm System. A system or group of systems in which the operations of circuits and devices are transmitted automatically to, recorded in, maintained by, and supervised from a listed central station that has competent and experienced servers and operators who, upon receipt of a signal, take such action as required by NFPA 72. (See Figure C.4.) *Note: Such service is to be controlled and operated by a person, firm, or corporation whose business is the furnishing, maintaining, or monitoring of supervised fire alarm systems.* NFPA 72, 2002 ed.

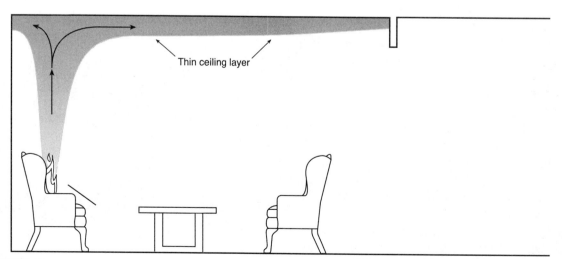

Figure C.3 Ceiling Layer.

Figure C.4 Central Station.

Certificate of Fitness. A written document issued by the authority having jurisdiction to any person for the purpose of granting permission to such person to conduct or engage in any operation or act for which certification is required. NFPA 1, 2003 ed.

Certification. Attests authoritatively; specifically, the issuance of a document that states that one has demonstrated the knowledge and skills necessary to function in a field. NFPA 1000, 2000 ed.

Certification/Certified. A system whereby a certification organization determines that a manufacturer has demonstrated the ability to produce a product that complies with the requirements of a standard, authorizes the manufacturer to use a label on listed products that comply with the requirements of the standard, and establishes a follow-up program conducted by the certification organization as a check of the methods the manufacturer uses to determine continued compliance with the requirements of the standard. NFPA 1971, 2000 ed.

Certification Mark or Label. The authorized identification symbol or logo of the certification organization. NFPA 1975, 2004 ed.

Certification Organization. An independent, third-party organization that determines product compliance with the requirements of a standard with a labeling/listing/follow-up program. NFPA 1971, 2000 ed.

Certifying Entity. An organization that is accredited to award certification to individuals. NFPA 1000, 2000 ed.

CFR. The Code of Federal Regulations of the United States Government. NFPA 1, 2003 ed.

CGA. The Compressed Gas Association. NFPA 55, 2005 ed.

Channel Access Time. The time elapsed from radio push-to-talk (PTT) to the time the receiving unit's speaker emits audio. NFPA 1221, 2002 ed.

Char. Carbonaceous material that has been burned and has a blackened appearance. NFPA 921, 2004 ed.

Char Blisters. Convex segments of carbonized material separated by cracks or crevasses that form on the surface of char, forming on materials such as wood as the result of pyrolysis or burning. NFPA 921, 2004 ed.

Characteristics. The physical science of fire, which includes the components of fire, the stages of fire development, heat transfer, smoke travel, and flame spread. NFPA 1035, 2005 ed.

Chassis. The basic operating motor vehicle including the engine, frame, and other essential structural and mechanical parts, but exclusive of the body and all appurtenances for the accommodation of driver, property, passengers, appliances, or equipment related to other than control. *Note: Common usage might, but need not, include a cab (or cowl).* NFPA 1901, 2003 ed.

Check Wood Irregularity. A separation of the wood along the fiber direction that usually extends across the rings of annual growth and commonly results from stresses set up in the wood during seasoning. NFPA 1931, 2004 ed.

Check-in. The process whereby resources first report to an incident. NFPA 1051, 2002 ed.

Chemical. Any compound, mixture, or solution in the form of a solid, liquid, or gas that may be hazardous by virtue of its properties other than or in addition to flammability or by virtue of the properties of compounds that might be evolved from hot work or cold work. NFPA 306, 2003 ed.

Chemical and Biological Terrorism Incidents. Situations involving the release of chemical or biological warfare agents in civilian areas by terrorists. NFPA 1991, 2005 ed.

Chemical Flash Fire. The ignition of a flammable and ignitible vapor or gas that produces an outward expanding flame front as those vapors or gases burn. *Note: This burning and expanding flame front, a fireball, will release both thermal and kinetic energy to the environment.* NFPA 1991, 2005 ed.

Chemical Heat of Combustion (Hc). The amount of heat released, in kJ/g (Btu/lb), when a substance is oxidized to yield stable end products, including water as a vapor, as measured under actual fire conditions in a normal ambient (air) atmosphere. NFPA 30B, 2002 ed.

Chemical Name. The scientific designation of a chemical in accordance with the nomenclature system developed by the International Union of Pure and Applied Chemistry or the Chemical Abstracts Service rules of nomenclature, or a name that clearly identifies a chemical for the purpose of conducting an evaluation. NFPA 1, 2003 ed.

Chemical Plant. A large integrated plant or that portion of such a plant, other than a refinery or distillery, where liquids are produced by chemical reactions or used in chemical reactions. NFPA 30, 2003 ed.

Chemical Terrorism Agents. Liquid, solid, gaseous, and vapor chemical agents and dual-use industrial chemicals used to inflict lethal or incapacitating casualties, generally on a civilian population as a result of a terrorist attack. NFPA 1994, 2001 ed.

Chemical Warfare Agent. A chemical substance intended to kill, seriously injure, or incapacitate humans through physiological effects. NFPA 1991, 2005 ed.

Chemical Warfare (CW) Agents. Liquid, solid, and gas chemical agents (most are liquids) traditionally used during warfare or armed conflict to kill or incapacitate an enemy. NFPA 1994, 2001 ed.

Chemical-Protective Layer. The material or composite used in an ensemble or clothing for the purpose of providing protection from chemical hazards. NFPA 1991, 2005 ed.

Chemical-Protective Material. Any material or composite used in liquid splash-protective ensemble or clothing for the purpose of providing protection from chemical hazards; can be a part of the "primary suit material." NFPA 1992, 2005 ed.

Chemical/Biological Barrier Material. The layer or part of the composite that is intended to provide a barrier of protection against chemical/biological terrorism agents. NFPA 1994, 2001 ed.

Chemical/Biological Terrorism Agents. Refers to chemical terrorism agents, biological terrorism agents, and dual-use industrial chemicals. NFPA 1994, 2001 ed.

Chemical/Biological Terrorism Incident Protective Ensembles. Multiple elements, categorized as Class 1, Class 2, or Class 3 ensembles, designed to provide minimum full body protection against exposure to chemical/biological terrorism agents occurring during chemical/biological terrorism emergencies. NFPA 1994, 2001 ed.

Chemical/Biological Terrorism Incident Protective Footwear. An element of the chemical/biological terrorism incident protective ensemble designed to provide minimum protection to the foot, ankle, and lower leg. NFPA 1994, 2001 ed.

Chemical/Biological Terrorism Incident Protective Garment(s). An element of the chemical/biological terrorism incident protective ensemble designed to provide minimum protection to the upper and lower torso, arms, and legs, excluding the head, hands, and feet. NFPA 1994, 2001 ed.

Chemical/Biological Terrorism Incident Protective Glove(s). An element of the chemical/biological terrorism incident protective ensemble designed to provide minimum protection to the wearer's hands and wrists. NFPA 1994, 2001 ed.

Chest Circumference. Measurement of upper torso garment from folded edge to folded edge, at base of armholes, and multiplied by 2 to obtain circumference. NFPA 1977, 2005 ed.

Chief Mate. The deck officer immediately responsible to the vessel's master. NFPA 1405, 2001 ed.

Chin Strap. A helmet term for the adjustable strap, fitting under the chin, to help secure the helmet to the head. NFPA 1971, 2000 ed.

Chocks. Fittings usually found on the rail or deck of a vessel having jaws that serve as fair leads for anchor rode and other lines. NFPA 1925, 2004 ed.

Circuit. The conductor, or radio channel, and associated equipment that are used to perform a specific function in connection with an alarm system. NFPA 1221, 2002 ed.

Claims Analyst. An internal or external person (depending on risk financing processes being used) expected to investigate the claim, evaluate it, prepare a position, ensure the appropriate "network" is involved, and, if necessary, begin negotiation of a settlement. NFPA 1250, 2004 ed.

Claims Made. The loss/occurrence and claim are made during the policy period. NFPA 1250, 2004 ed.

Claims Occurrence. The loss occurs during the policy period; the claim can be made at any time. NFPA 1250, 2004 ed.

Class 1 Ensemble. A chemical/biological terrorism incident protective ensemble to protect fire and emergency services personnel at chemical/biological terrorism incidents where the identity or concentration of the vapor or liquid agent is unknown, or where it is necessary to provide vapor protection, or where liquid contact is expected and no direct skin contact can be permitted as exposure of personnel at these levels will result in the substantial possibility of immediate death, immediate serious incapacitation, or the ability to escape will be severely impaired. NFPA 1994, 2001 ed.

Class 1 Leakage. Seepage of fluid, as indicated by wetness or discoloration, not great enough to form drops. NFPA 1915, 2000 ed.

Class 1 Unstable (Reactive) Material. A material that, in itself, is normally stable but can become unstable at elevated temperatures and pressures. NFPA 1, 2003 ed.

Class 1 Water-Reactive Material. Materials that can react with water with some release of energy, but not violently. NFPA 1, 2003 ed.

Class 2 Ensemble. A chemical/biological terrorism incident protective ensemble to protect fire and emergency services personnel at chemical/biological terrorism incidents where it is necessary to provide sufficient vapor protection for the intended operation, where direct contact of liquid droplets is probable, and where victims are not ambulatory but symptomatic. NFPA 1994, 2001 ed.

Class 2 Leakage. Leakage of fluid great enough to form drops, but not enough to cause drops to fall from the item being inspected. NFPA 1915, 2000 ed.

Class 2 Unstable (Reactive) Material. A material that readily undergoes violent chemical change at elevated temperatures and pressures. NFPA 1, 2003 ed.

Class 2 Water-Reactive Material. Materials that can form potentially explosive mixtures with water. NFPA 1, 2003 ed.

Class 3 Ensemble. A chemical/biological terrorism incident protective ensemble to protect fire and emergency services personnel at chemical/biological terrorism incidents where it is necessary to provide sufficient liquid protection for the intended operation, where direct contact of liquid droplets is possible, and where victims are not impaired but ambulatory. NFPA 1994, 2001 ed.

Class 3 Leakage. Leakage of fluid great enough to cause drops to fall from the item being inspected. NFPA 1915, 2000 ed.

Class 3 Unstable (Reactive) Material. A material that, in itself, is capable of detonation or explosive decomposition or explosive reaction but requires a strong initiating source or must be heated under confinement before initiation. NFPA 1, 2003 ed.

Class 3 Water-Reactive Material. Materials that react explosively with water without requiring heat or confinement. NFPA 1, 2003 ed.

Class 4 Unstable (Reactive) Material. A material that, in itself, is readily capable of detonation or explosive decomposition or explosive reaction at normal temperatures and pressures. NFPA 1, 2003 ed.

Class A Fire. A fire in ordinary combustible materials, such as wood, cloth, paper, rubber, and many plastics. (See Figure C.5.) NFPA 10, 2002 ed.

Figure C.5 Example of Class A Fire.

Class A Foam. Foam for use on fires in Class A fuels. NFPA 1150, 2004 ed.

Class A Fuel. Materials such as vegetation, wood, cloth, paper, rubber, and some plastics in which combustion can occur at or below the surface of the material. NFPA 1150, 2004 ed.

Class A Mercantile Occupancy. All mercantile occupancies having an aggregate gross area of more than 2800 m^2 (30,000 ft^2) or occupying more than three stories for sales purposes. NFPA 1, 2003 ed.

Class A Pyroxylin Products. Those products made from material over 1/10 in. (2.5 mm) thick and weighing over 0.5 oz (14 g) and that are not finely divided during manufacture into teeth, scrollwork, or projections. NFPA 42, 2002 ed.

Class B. Flammable liquids. NFPA 402, 2002 ed.

Class B Fire. A fire in flammable liquids, combustible liquids, petroleum greases, tars, oils, oil-based paints, solvents, lacquers, alcohols, and flammable gases. (See Figure C.6.) NFPA 10, 2002 ed.

Class B Foam. Foam intended for use on Class B fires. NFPA 1901, 2003 ed.

Class B Mercantile Occupancy. All mercantile occupancies of more than 280 m^2 (23000 ft^2), but not more than 2800 m^2 (30,000 ft^2), aggregate gross area and occupying not more than three stories for sales purposes. NFPA 1, 2003 ed.

Class C Fire. A fire that involves energized electrical equipment. (See Figure C.7.) NFPA 10, 2002 ed.

Class C Mercantile Occupancy. All mercantile occupancies of not more than 280 m^2 (3000 ft^2) gross area and used for sales purposes occupying one story only. NFPA 1, 2003 ed.

FIGURE C.6 Example of Class B Fire.

Class D. Combustible metals. NFPA 402, 2002 ed.

Class D Fire. Fire in combustible metals, such as magnesium, titanium, zirconium, sodium, lithium, and potassium. NFPA 10, 2002 ed.

Clean Burn. A fire pattern on surfaces where soot has been burned away. (See Figure C.8.) NFPA 921, 2004 ed.

Clean Zone. A defined space in which the concentration of airborne particles is controlled to specified limits. NFPA 318, 2002 ed.

Cleaning. The physical removal of dirt and debris, which generally is accomplished with soap and water and physical scrubbing. NFPA 1581, 2005 ed.

FIGURE C.7 Example of Class C Fire.

FIGURE C.8 Example of a Clean Burn.

Cleaning Glove. Multipurpose glove, not intended for emergency patient care, that provides a barrier against body fluids, cleaning fluids, and disinfectants and limited physical protection to the wearer during cleaning or care of emergency medical clothing and equipment. NFPA 1999, 2003 ed.

Clear Text. The use of plain language in radio communications transmissions. NFPA 1561, 2005 ed.

Cleat. Fitting attached to the vessel used to secure an anchor rode or other line to the vessel. NFPA 1925, 2004 ed.

Close-off Pressure. The maximum pressure the pump is capable of developing at zero discharge flow. NFPA 1925, 2004 ed.

Closed-Circuit Self-Contained Breathing Apparatus (SCBA). A recirculation-type SCBA in which the exhaled gas is re-breathed by the wearer after the carbon dioxide has been removed from the exhalation gas and the oxygen content within the system has been restored from sources such as compressed breathing air, chemical oxygen, liquid oxygen, or compressed gaseous oxygen. (See Figure C.9.) NFPA 1981, 2002 ed.

Closed Container. A container as herein defined, so sealed by means of a lid or other device that neither liquid nor vapor will escape from it at ordinary temperatures. NFPA 30A, 2003 ed.

Figure C.9 Closed-Circuit SCBA.

Closed System Use. Use of a solid or liquid hazardous material in a closed vessel or system that remains closed during normal operations where vapors emitted by the product are not liberated outside of the vessel or system and the product is not exposed to the atmosphere during normal operations and all uses of compressed gases. NFPA 1, 2003 ed.

Coaming. The raised framework around deck or bulkhead openings to prevent entry of water. NFPA 1405, 2001 ed.

Coat. A protective garment that is an element of the protective ensemble designed to provide minimum protection to upper torso and arms, excluding the hands and head. NFPA 1976, 2000 ed.

Coating. *As applied to aircraft:* Application of special-purpose materials such as anticorrosion and walkway paints. NFPA 410, 2004 ed. *As applied to fire hose:* A protective material impregnated, saturated, or coated on the outside reinforcement layer of the hose to provide additional reinforcement or protection for the hose. NFPA 1962, 2003 ed.

Coaxial Cable. A transmission line in which one conductor completely surrounds the other, the two being coaxial and separated by a continuous solid dielectric or by dielectric spacers. NFPA 1221, 2002 ed.

Cockpit Voice Recorder (CVR). A device that monitors flight deck crew communications through a pickup on the flight deck connected to a recorder that is usually mounted in the tail area of the aircraft and that is designed to withstand certain impact forces and a degree of fire. NFPA 402, 2002 ed.

Code. A standard that is an extensive compilation of provisions covering broad subject matter or that is suitable for adoption into law independently of other codes and standards. NFPA Official Definition.

Cofferdam. A void between compartments or tanks of a ship for purposes of isolation. NFPA 1405, 2001 ed.

Cold Zone. *As applied to hazardous materials:* The control zone of a hazardous materials incident that contains the command post and such other support functions as are deemed necessary to control the incident. NFPA 472, 2002 ed. *As applied to airport fire fighting:* The hazard-free area around an incident. NFPA 1003, 2005 ed.

Collapse Support Operations. Operations performed at the scene that include providing for rescuer comfort, scene lighting, scene management, and equipment readiness. NFPA 1006, 2003 ed.

Collapse Type. Five general types of collapse include lean-to collapse, "V" shape collapse, pancake collapse, cantilever collapse, and A-frame collapse. NFPA 1006, 2003 ed.

Collar. The portion of a coat or coverall that encircles the neck. NFPA 1971, 2000 ed.

Collar Length. Upper torso garment measurement along top of collar from point-to-point. NFPA 1977, 2005 ed.

Collar Lining. The part of the collar fabric composite that is next to the skin when the collar is closed in the raised position. NFPA 1971, 1997 ed.

Collar Width. Upper torso garment measurement at center back from top edge of unfolded collar to the bottom collar seam. NFPA 1977, 2005 ed.

Collector Rings. An assembly of slip rings for transferring electrical energy from a stationary to a rotating member. NFPA 70, 2005 ed.

Combi. An aircraft designed to transport both passengers and cargo on the same level within the fuselage. NFPA 402, 2002 ed.

Combination Fire Apparatus. A vehicle consisting of a pulling tractor and trailer. NFPA 1915, 2000 ed.

Combination Fire Department. A fire department having emergency service personnel comprising less than 85 percent majority of either volunteer or career membership. NFPA 1720, 2004 ed.

Combination Ladder. A ground ladder that is capable of being used both as a stepladder and as a single or extension ladder. (See Figure C.10.) NFPA 1932, 2004 ed.

Combination SCBA/SAR. An atmosphere-supplying respirator that supplies a respirable atmosphere to the user from a combination of two breathing air sources that both are independent of the ambient environment and consists of (1) an open-circuit SCBA certified as compliant with NFPA 1981, Standard on Open-Circuit Self-Contained Breathing Apparatus for Fire and Emergency Services, and having a minimum rated service time of 30 minutes; and (2) having a connection for the attachment of an air line that would provide a continuous supply of breathing air independent of the SCBA breathing air supply. NFPA 1981, 2002 ed.

Combination Tool. A powered rescue tool that is capable of at least spreading and cutting. NFPA 1936, 2005 ed.

Combination Vehicle. A vehicle consisting of a pulling tractor and trailer. NFPA 1901, 2003 ed.

Combustible. Capable of reacting with oxygen and burning if ignited. NFPA 220, 2005 ed.

Combustible Dust. Any finely divided solid material that is 420 microns or smaller in diameter (material passing a U.S. No. 40 Standard Sieve) and presents a fire or explosion hazard when dispersed and ignited in air. NFPA 654, 2000 ed.

Combustible Fiber. Any material in a fibrous or shredded form that will readily ignite when heat sources are present. NFPA 1, 2003 ed.

Combustible Gas Indicator. An instrument that samples air and indicates whether there are ignitable vapors or gases present. NFPA 921, 2004 ed.

Combustible Liquid. Any liquid that has a closed-cup flash point at or above 37.8°C (100°F). NFPA 306, 2003 ed.

Combustible Refuse. All combustible or loose rubbish, litter, or waste materials generated by an occupancy that are refused, rejected, or considered worthless and are disposed of by incineration on the premises where generated or periodically transported from the premises. NFPA 1, 2003 ed.

Combustible Waste. Combustible or loose waste material that is generated by an establishment or process and, if salvageable, is retained for scrap or reprocessing on the premises where generated or transported to a plant for processing. NFPA 1, 2003 ed.

Combustion. A chemical process of oxidation that occurs at a rate fast enough to produce heat and usually light in the form of either a glow or flame. NFPA 5000, 2005 ed.

Combustion Products. Constituents resulting from the combustion of a fuel with the oxygen in the air, including the inert but excluding excess air. NFPA 54, 2002 ed.

Command. The act of directing and/or controlling resources by virtue of explicit legal, agency, or delegated authority. NFPA 1143, 2003 ed.

Command and Communications Apparatus. A fire apparatus used primarily for communications and incident command. NFPA 1901, 2003 ed.

Command Staff. Positions that are established to assume responsibility for key activities in the incident management system that are not a part of the line organization that include safety officer, information officer, and liaison officer. NFPA 1561, 2005 ed.

FIGURE C.10 Combination Ladder.

Commercial and Truck Repair Garages. Buildings, structures, or portions thereof used for the storage, maintenance, and repair of commercial motor vehicles or trucks, including fleets of motor vehicles operated by utilities, large businesses, mercantile, rental agencies, and other similar concerns. *Note: Facilities for the dispensing of motor fuels are commonly provided in connection with these garages.* NFPA 1, 2000 ed.

Common Battery. The battery used to power recorders, transmitters, relays, other communications center equipment, and subsidiary communications center equipment. NFPA 1221, 2002 ed.

Common Path of Travel. The portion of exit access that must be traversed before two separate and distinct paths of travel to two exits are available. NFPA 101, 2003 ed.

Communicable Disease. A disease that can be transmitted from one person to another. NFPA 1500, 2002 ed.

Communications Center. A building or portion of a building that is specifically configured for the primary purpose of providing emergency communications services or public safety answering point (PSAP) services to one or more public safety agencies under the authority or authorities having jurisdiction. NFPA 1221, 2002 ed.

Communications Officer/Unit Leader. The individual responsible for development of plans to make the most effective use of incident-assigned communications equipment and facilities, installation and testing of all communications equipment, supervision and operation of the incident communications center, distribution and recovery of equipment assigned to incident personnel, and maintenance and on-site repair of communications equipment. NFPA 1221, 2002 ed.

Communications System. A combination of links or networks that serve a general function such as a system made up of command, tactical, logistical, and administrative networks. NFPA 1221, 2002 ed.

Community Resource List. A list that includes all private and public contact numbers that will provide the available community resources to mitigate a specified type or range of rescue incidents and hazardous conditions in the community. *Note: A form of agreement or contract negotiated prior to the potential incident with participating concerns will enhance reliability of the resources.* NFPA 1006, 2003 ed.

Companionway. An interior stair-ladder used to travel from deck to deck, usually enclosed. NFPA 1405, 2001 ed.

Company. A group of members: (1) Under the direct supervision of an officer; (2) Trained and equipped to perform assigned tasks; (3) Usually organized and identified as engine companies, ladder companies, rescue companies, squad companies, or multi-functional companies; (4) Operating with one piece of fire apparatus (engine, ladder truck, elevating platform, quint, rescue, squad, ambulance) except where multiple apparatus are assigned that are dispatched and arrive together, continuously operate together, and are managed by a single company officer; (5) Arriving at the incident scene on fire apparatus. NFPA 1500, 2002 ed.

Company Officer. The officer or any other position of comparable responsibility in the department in charge of a fire department company or station. NFPA 1143, 2003 ed.

Compass. A device that uses the earth's magnetic field to indicate relative direction. NFPA 1670, 2004 ed.

Compatible Material. A material that, when in contact with an oxidizer, will not react with the oxidizer or promote or initiate its decomposition. NFPA 1, 2003 ed.

Competent Person. One who is capable of identifying existing and predictable hazards in the surroundings or working conditions that are unsanitary, hazardous, or dangerous to employees, and who has authorization to take prompt corrective measures to eliminate them. NFPA 1006, 2003 ed.

Compliance/Compliant. Meeting or exceeding all applicable requirements of a standard. NFPA 1971, 2000 ed.

Compliant Product. Product that is covered by a standard and has been certified as meeting all applicable requirements of the standard that pertain to the product. NFPA 1999, 2003 ed.

Component. *As applied to protective ensembles:* Any material, part, or subassembly used in the construction of the protective ensemble or any element of the protective ensemble. NFPA 1971, 2000 ed. *As applied to fire and emergency services:* Any material, part, or subassembly providing the required protection that is used in the construction of the SCBA. NFPA 1981, 2002 ed. *As applied to fire apparatus maintenance:* A constituent part of a mechanical or electrical device. NFPA 1915, 2000 ed. *As applied to fire apparatus refurbishing:* A constituent part of a fire apparatus or system. NFPA 1912, 2001 ed.

Component Part(s). Any material(s) or part(s) used in the construction of a vapor-protective ensemble or ensemble elements. NFPA 1991, 2005 ed.

Components. All materials and hardware—such as emblems, thread, trim, bindings, zippers, snaps, buttons, and labels, but excluding interlinings—used in the construction of station/work uniforms. NFPA 1975, 1999 ed.

Components of Emergency Medical Service (EMS) System. The parts of a comprehensive plan to treat an individual in need of emergency medical care following an illness or injury. NFPA 473, 2002 ed.

Composite Materials. Lightweight materials having great structural strength. *Note: They are made of fine fibers embedded in carbon/epoxy materials. The fibers are usually boron, fiberglass, aramid, or carbon in the form of graphite. Composite materials do not present unusual fire fighting problems, but products of their combustion should be considered a respiratory hazard to fire fighters.* NFPA 402, 2002 ed.

Compound Gauge. A gauge that indicates pressure both above and below atmospheric pressure. NFPA 1911, 2002 ed.

Compound Rope Mechanical Advantage System. A combination of individual rope mechanical advantage systems created by stacking the load end of one rope mechanical advantage system onto the haul line of another or others to multiply the forces created by the individual system(s). NFPA 1670, 1999 ed.

Compressed Air Foam System (CAFS). A foam system that combines air under pressure with foam solution to create foam. NFPA 1901, 2003 ed.

Compressed Breathing Air. Oxygen or a respirable gas mixture stored in a compressed state and supplied to the user in gaseous form. NFPA 1981, 2002 ed.

Compressed Gas Mixtures. A mixture of two or more compressed gases contained in a packaging, the hazard properties of which are represented by the properties of the mixture as a whole. NFPA 1, 2003 ed.

Compressed Gas System. An assembly of equipment designed to contain, distribute, or transport compressed gases. NFPA 318, 2002 ed.

Compressed Gases in Solution. Nonliquefied gases that are dissolved in a solvent. NFPA 1, 2003 ed.

Computer. A programmable electronic device that contains a central processing unit(s), main storage, an arithmetic unit, and special register groups. NFPA 1221, 2002 ed.

Concentration. The ratio of the amount of one constituent of a homogenous mixture to the total amount of all constituents in the mixture. NFPA 53, 2004 ed.

Condensate (Condensation). The liquid that separates from a gas (including flue gas) due to a reduction in temperature or an increase in pressure. NFPA 54, 2002 ed.

Conduction. Heat transfer to another body or within a body by direct contact. (See Figure C.11.) NFPA 921, 2004 ed.

Confidentiality. A principle of law and professional ethics that recognizes the privacy of individuals. NFPA 1035, 2005 ed.

Confine a Fire. To restrict the fire within determined boundaries established either prior to the fire or during the fire. NFPA 901, 2001 ed.

Confined Space. An area large enough and so configured that a member can bodily enter and perform assigned work but that has limited or restricted means for entry and exit and is not designed for continuous human occupancy. NFPA 1500, 2002 ed.

Confined Space Approach. The means of approach to the entry opening of a confined space. NFPA 1006, 2003 ed.

Confined Space Entry. Includes ensuing work activities in a confined space and is considered to have occurred as soon as any part of the entrant's body breaks the plane of an opening into the space. NFPA 1006, 2003 ed.

Confined Space Entry Opening. The port or opening used to enter a confined space. NFPA 1006, 2003 ed.

Confined Space Entry Permit. A written or printed document established by an employer, in applicable U.S. federally regulated industrial facilities for nonrescue entry into confined spaces, that authorizes specific employees to enter a confined space and contains specific information as required. NFPA 1006, 2003 ed.

Confined Space Rescue Preplan. An informational document completed by rescue personnel pertaining to a specific space. *Note: The document should include but is not limited to information concerning hazard abatement requirements, access to the space, size and type of entry openings, internal configuration of the space, and a suggested action plan for rescue of persons injured within the space.* NFPA 1006, 2003 ed.

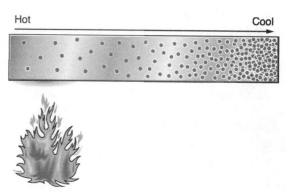

FIGURE C.11 Conduction.

Confined Space Rescue Service. The confined space rescue team designated by the AHJ to rescue victims from within confined spaces, including operational and technical levels of industrial, municipal, and private sector organizations. NFPA 1670, 2004 ed.

Confined Space Rescue Team. A combination of individuals trained, equipped, and available to respond to confined space emergencies. (See Figure C.12.) NFPA 1670, 2004 ed.

Confined Space Type. A classification of confined spaces that incorporates the size, configuration, and accessibility of an entry opening as well as the internal configuration/entanglement structures within the space. NFPA 1006, 2003 ed.

Confinement. Those procedures taken to keep a material, once released, in a defined or local area. NFPA 471, 2002 ed.

Constant Gallonage Spray Nozzle. An adjustable-pattern spray nozzle that discharges a constant discharge rate throughout the range of patterns from a straight stream to a wide spray at a designed nozzle pressure. NFPA 1964, 2003 ed.

Constant Pressure (Automatic) Spray Nozzle. An adjustable-pattern spray nozzle in which the pressure remains relatively constant through a range of discharge rates. NFPA 1964, 2003 ed.

Constant/Select Gallonage Spray Nozzle. A constant discharge rate spray nozzle with a feature that allows manual adjustment of the orifice to effect a predetermined discharge rate while the nozzle is flowing. NFPA 1964, 2003 ed.

Construction Classification Number. A series of numbers from 0.5 through 1.5 that are mathematical factors used in a formula to determine the total water supply requirements. NFPA 1142, 2001 ed.

Construction Documents. Documents that consist of scaled design drawings and specifications for the purpose of construction of new facilities or modification to existing facilities. NFPA 1, 2003 ed.

Construction Grade Lumber. Lumber products that are readily available in sizes and lengths for general construction applications. NFPA 1006, 2003 ed.

Construction Type. The combination of materials used in the construction of a building or structure, based on the varying degrees of fire resistance and combustibility. NFPA 5000, 2002 ed.

Consumer Fireworks. Any small fireworks device designed primarily to produce visible effects by combustion or deflagration that complies with the construction, chemical composition, and labeling regulations of the U.S. Consumer Product Safety Commission, as set forth in 16 CFR 1500 and 1507. NFPA 1123, 2000 ed.

Consumer Fireworks Retail Sales Area. The portion of a consumer fireworks retail sales facility or store, including the immediately adjacent aisles, where consumer fireworks are located for the purpose of retail display and sale to the public. NFPA 1124, 2003 ed.

Consumer Fireworks Retail Sales Facility. A permanent or temporary building or structure, consumer fireworks retail sales stand, tent, canopy, or membrane structure that is used primarily for the retail display and sale of consumer fireworks to the public. NFPA 1124, 2003 ed.

Consumer Fireworks Retail Sales Stand. A temporary or permanent building or structure that has a floor area not greater than 74 m^2 (800 ft^2), other than tents, canopies, or membrane structures, that is used primarily for the retail display and sale of consumer fireworks to the public. NFPA 1124, 2003 ed.

Contain. To take suppression action that can reasonably be expected to check the fire spread under prevailing and predicted conditions. NFPA 1051, 2002 ed.

FIGURE C.12 Confined Space Rescue Team.

Contain a Fire. To take suppression action as needed that can reasonably be expected to check the fire spread under prevailing conditions. NFPA 901, 2001 ed.

Container Terminal. A terminal that is designed to handle containers that are carried by truck or rail car when transported over land. NFPA 1405, 2001 ed.

Containment. The actions taken to keep a material in its container (e.g., stop a release of the material or reduce the amount being released). NFPA 471, 2002 ed.

Contaminant. A harmful, irritating, or nuisance material foreign to the normal atmosphere. NFPA 1404, 2002 ed.

Contaminated. The presence or the reasonably anticipated presence of blood, body fluids, or other potentially infectious materials on an item or surface. NFPA 1581, 2005 ed.

Contaminated Sharps. Any contaminated object that can penetrate the skin including, but not limited to, needles, lancets, scalpels, broken glass, jagged metal, or other debris. NFPA 1581, 2005 ed.

Contamination. The process of transferring a hazardous material from its source to people, animals, the environment, or equipment, which may act as a carrier. NFPA 471, 2002 ed.

Contamination/Contaminated. The process by which ensembles and ensemble elements are exposed to hazardous materials or biological agents. NFPA 1851, 2001 ed.

Continual. With respect to the testing of a powered rescue tool, a test sequence performed with pauses or interruptions. NFPA 1936, 2005 ed.

Continuous. With respect to the testing of a powered rescue tool, a test sequence performed without any pauses or interruptions for any purpose. NFPA 1936, 2005 ed.

Continuous Egress. A continuous exit or rescue path down an aerial device from an elevated position to the ground. NFPA 1901, 2003 ed.

Contract Cleaning. Cleaning conducted by a facility outside the organization that specializes in cleaning protective clothing. NFPA 1851, 2001 ed.

Contractor. The person or company responsible for fulfilling an agreed-upon contract. NFPA 1901, 2003 ed.

Control. The point in time when the perimeter spread of a wildland fire has been halted and can be reasonably expected to hold under foreseeable conditions. NFPA 1051, 2002 ed.

Control a Fire. To complete a control line around a fire, any spot fire therefrom, or any interior island to be saved; to burn out any unburned area adjacent to the fire side of the control line and to cool down all hot spots that are an immediate threat to the control line. NFPA 1143, 2003 ed.

Control Combustion Process. Control the inherent fire behavior. NFPA 550, 2002 ed.

Control Console. A system containing controls to operate communications equipment. NFPA 1221, 2002 ed.

Control Line. All constructed or natural barriers and the treated fire edge used to control a fire. NFPA 1051, 2002 ed.

Control Zones. The areas at hazardous materials incidents that are designated based upon safety and the degree of hazard. (See Figure C.13.) NFPA 471, 2002 ed.

Controlled Atmosphere (CA) Warehouse. A facility for storing specialty products, such as fruits, that generally includes sealed storage rooms, with controlled temperature and air content, the most common being an atmosphere containing a high percentage of a gas such as nitrogen. NFPA 1620, 2003 ed.

Convection. Heat transfer by circulation within a medium such as a gas or a liquid. (See Figure C.14.) NFPA 921, 2004 ed.

Convenient Reach. *As applied to fire fighting:* The ability of the operator to manipulate the controls from a driving/riding position without excessive movement away from the seat back or without excessive loss of eye contact with the roadway. NFPA 1901, 1999 ed. *As applied to marine vessels:* In marine fire fighting vessels, the ability to operate controls without excessive movement from a fixed position such as a seat or safety harness. NFPA 1925, 2004 ed.

Conventional Pallets. A material-handling aid designed to support a unit load with openings to provide access for material-handling devices. NFPA 230, 2003 ed.

Conventional Radio. A radio system in which automatic computer control of channel assignments is not required or used, system-managed queuing of calls is not provided, and channels are selected manually by the users. NFPA 1221, 2002 ed.

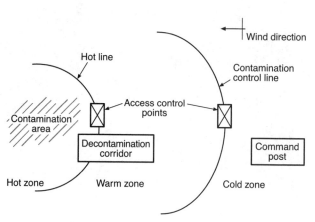

Figure C.13 Control Zones.

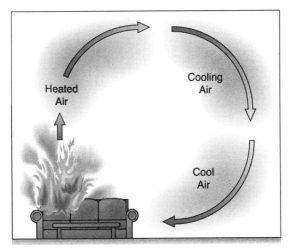

FIGURE C.14 Convection.

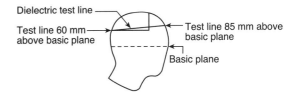

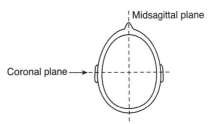

FIGURE C.15 Coronal Plane.

Cooking Fire. The noncommercial, residential burning of materials not exceeding 3 ft (0.9 m) in diameter and 2 ft (0.6 m) in height, other than rubbish in which the fuel burned is contained in an outdoor fireplace, a barbecue grill, or a barbecue pit for the purpose of preparing food. NFPA 1, 2003 ed.

Cooling Preheater Device. A device for heating the engine coolant so that the engine is maintained at a constant temperature. NFPA 414, 2001 ed.

Coordinated Universal Time (UTC). A coordinated time scale, maintained by the Bureau Internationale des Poids et Mesures (BIPM), which forms the basis of a coordinated dissemination of standard frequencies and time signals. NFPA 1221, 2002 ed.

Coordination. The process used to get people, who could represent different agencies, to work together integrally and harmoniously in a common action or effort. NFPA 472, 2002 ed.

Coronal Plane. The anatomical plane perpendicular to both the basic and midsagittal planes and containing the midpoint of a line connecting the superior rims of the right and left auditory meatuses. (See Figure C.15.) NFPA 1976, 2000 ed.

Corrective Lens. A lens designed to fit the specifications of the wearer's individual corrective prescription. NFPA 1404, 2002 ed.

Corrosion. A condition exhibiting any signs of deterioration, including pitting or loss of metal. NFPA 1983, 2001 ed.

Corrosive. A chemical that causes visible destruction of, or irreversible alterations in, materials by chemical action at the site of contact. NFPA 1, 2003 ed.

Corrosive Gas. A gas that causes visible destruction of or irreversible alterations in living tissue by chemical action at the site of contact. NFPA 55, 2005 ed.

COTP. United States Coast Guard Captain of the Port. NFPA 1405, 2001 ed.

Counter Balance. A raising system utilizing a 1:1 mechanical advantage and a weighted object (human or otherwise) to reduce the need for additional force to lift the load. NFPA 1670, 1999 ed.

Coupling Assembly. A complete coupling including its gaskets and the expansion rings or collar pieces used in attaching the coupling to the hose. NFPA 1963, 1998 ed.

Couplings. One set or pair of connection devices attached to a fire hose that allow the hose to be interconnected to additional lengths of hose or adapters and other fire fighting appliances. (See Figure C.16.) NFPA 1963, 1998 ed.

Court. An open, uncovered, unoccupied space, unobstructed to the sky, bounded on three or more sides by exterior building walls. NFPA 101, 2003 ed.

Cover. An additional layer on the outside of a hose consisting of a continuous synthetic rubber or plastic that is usually thicker than a coating. NFPA 1961, 2002 ed.

Coverall. A protective garment that is an element of the protective ensemble configured as a single-piece garment and designed to provide minimum protection to the torso, arms, and legs, excluding the head, hands, and feet. NFPA 1976, 2000 ed.

Covered Fuse. A fuse that is protected against accidental ignition by contact with a spark, smoldering item, or small open flame. NFPA 1124, 2003 ed.

FIGURE C.16 Couplings.

Covered Hose. A hose with a jacket covered and lined with a continuous synthetic rubber or plastic. The cover is usually thicker than a coating. (See Figure C.17.) NFPA 1962, 2003 ed.

Covered Mall Building. A building, including the covered mall, enclosing a number of tenants and occupancies, such as retail stores, drinking and dining establishments, entertainment and amusement facilities, offices, and other similar uses, wherein two or more tenants have a main entrance into the covered mall. NFPA 101B, 2002 ed.

Covered Pier. A fixed or floating pier that is provided with a roof system to protect berthed boats from the weather. NFPA 303, 2000 ed.

Covered Storage. A structure or building capable of receiving and storing boats for extended periods of time while protecting the boats from exposure to the weather. NFPA 303, 2000 ed.

Cracking Pressure. The pressure at which the suit exhaust valve begins to open, releasing exhaust air to the outside suit environment. NFPA 1991, 2005 ed.

Craze. The appearance of fine cracks in surface of helmet shell or other smooth surface of an element. NFPA 1851, 2001 ed.

Create. Design original educational or informational resource materials. NFPA 1035, 2005 ed.

Creep. Unintended movement. NFPA 1936, 2005 ed.

Crew. *As applied to marine vessels:* Anyone associated with the operation of the vessel. NFPA 1925, 2004 ed. *As applied to fire fighting:* A two-person team of fire fighters. (See Figure C.18.) NFPA 1500, 2002 ed.

Crew Boss (Leader). A person who is in supervisory charge of usually 10 to 20 fire fighters and who is responsible for their performance, safety, and welfare. NFPA 1143, 2003 ed.

Cribbing. Short lengths of timber/composite materials, usually 101.60 mm × 101.60 mm (4 in. × 4 in.) and 457.20 mm × 609.60 mm (18 in. × 24 in.) long that are used in various configurations to stabilize loads in place or while load is moving. NFPA 1006, 2003 ed.

Critical Angle. An angle of 120 degrees or less created between two rope rescue system components wide enough so as to create excessive force on the anchor points to which they are attached. NFPA 1670, 2004 ed.

Critical Area. The area calculated to be one-half the overall length of the helicopter multiplied by three times the width of the widest portion of the fuselage. NFPA 418, 2001 ed.

FIGURE C.17 Rubber-Covered Hose.

FIGURE C.18 Two-Person Crew.

Critical Incident Stress Debriefing (CISD). A postincident meeting designed to assist rescue personnel in dealing with psychological trauma as the result of an emergency. (See Figure C.19.) NFPA 1006, 2003 ed.

Critical Incident Stress Management (CISM). A program designed to reduce acute and chronic effects of stress related to job functions. NFPA 450, 2004 ed.

Critical Radiant Flux. The level of incident radiant heat energy on the floor covering system at the most distant flameout point. *Note: Critical radiant flux is reported as W/cm². NFPA 253, 2000 ed.*

Critical Rescue and Fire Fighting Access Area (CRFFAA). The rectangular area that surrounds a runway within which aircraft movements can be expected to occur on airports and whose width extends 150 m (500 ft) from each side of the runway centerline and whose length is 1000 m (3300 ft) beyond each runway threshold. NFPA 405, 2004 ed.

Critique. A postincident analysis of the effectiveness of the rescue effort. NFPA 1006, 2003 ed.

Cross Braces (or Struts). The individual horizontal members of a shoring system installed perpendicular to the sides of the excavation, the ends of which bear against either uprights or wales. NFPA 1670, 1999 ed.

Cross Contamination. The transfer of contamination from one item to another or to the environment. NFPA 1852, 2002 ed.

Crown. The portion of the helmet that covers the head above the reference plane. NFPA 1971, 2000 ed.

Crown Straps. The part of the helmet suspension that passes over the head. NFPA 1971, 2000 ed.

Crude Petroleum. Hydrocarbon mixtures that have a flash point below 65.6°C (150°F) and that have not been processed in a refinery. NFPA 30, 2003 ed.

Figure C.19 CISD.

Crush Syndrome. A condition in which muscle death occurs because of pressure applied by an external load (e.g., a vehicle, parts of a fallen building, a rock, or a squeeze in a tight hole). NFPA 1006, 2003 ed.

Cryogenic Fluids Container. A cryogenic vessel used for transportation, handling, or storage. NFPA 1, 2003 ed.

Cryogenic Gas. A refrigerated liquid gas having a boiling point below –130°F (–90°C) at atmospheric pressure. NFPA 1994, 2001 ed.

Cuff. Finished edge of sleeve openings of protective garments. NFPA 1977, 2005 ed.

Cuff Circumference. Measurement of torso garment cuff along bottom of opening from folded edge to folded edge, and multiplied by 2 to obtain circumference. NFPA 1977, 2005 ed.

Curb Cut. Reduced curb height to facilitate vehicle passage over or across a curb. *Note: Curb cut can be an abrupt reduction or a tapering reduction for the length of the curb on each side of the means of access.* NFPA 1141, 2003 ed.

Current. A flow of electric charge. NFPA 921, 2004 ed.

Cut Sheet. A document that specifies the dimensions, slope, and other pertinent information regarding a particular excavation. NFPA 1006, 2003 ed.

Cut Station. A functional area or sector that utilizes lumber, timber, and an assortment of hand and power tools to complete operational objectives for stabilizing or shoring at a rescue incident or training evolution. NFPA 1006, 2003 ed.

Cutoff Room. A room within a building and having at least one exterior wall. NFPA 30, 2003 ed.

Cutter. A powered rescue tool with at least one movable blade that is used to cut, shear, or sever material. (See Figure C.20.) NFPA 1936, 2005 ed.

Cylinder. A portable compressed gas container fabricated to, or authorized for use by, the U.S. Department of Transportation (DOT), or fabricated to Transport Canada (TC) or the ASME Boiler and Pressure Vessel Code, Section VIII, "Rules for the Construction of Unfired Pressure Vessels." NFPA 560, 2002 ed.

Cylinder Containment System. A gastight recovery system comprised of equipment or devices that can be placed over a leak in a compressed gas container, thereby stopping or controlling the escape of gas from the leaking container. NFPA 55, 2005 ed.

FIGURE C.20 Cutter.

Cylinder Containment Vessel. A gastight recovery vessel designed so that a leaking compressed gas container can be placed within its confines, thereby encapsulating the leaking container. NFPA 55, 2005 ed.

Cylinder Links. The mechanisms that can be used in connecting an articulating boom to the end of the upper elevating cylinders or to the lower and upper booms. NFPA 1914, 2002 ed.

D

Damage Assessment. An appraisal or determination of the effects of the disaster on human, physical, economic, and natural resources. NFPA 1600, 2004 ed.

Damage Control Locker/Emergency Gear Locker. A locker used for the storage of emergency equipment. NFPA 1405, 2001 ed.

Dangerous Goods. Includes explosives and any other article defined as a combustible liquid, corrosive material, infectious substances, flammable compressed gases, oxidizing materials, poisonous articles, radioactive materials, and other restrictive articles. *Note: This term* dangerous goods *is synonymous with the terms* hazardous materials *and* restricted articles. *The term is used internationally in the transportation industry.* NFPA 402, 2002 ed.

Data Communication Channel. A single path for transmitting electric signals that is distinct from other parallel paths. NFPA 1221, 2002 ed.

Day-Care Home. A building or portion of a building in which more than three but not more than 12 clients receive care, maintenance, and supervision, by other than their relative(s) or legal guardians(s), for less than 24 hours per day. NFPA 101, 2003 ed.

Day-Care Occupancy. An occupancy in which four or more clients receive care, maintenance, and supervision, by other than their relatives or legal guardians, for less than 24 hours per day. NFPA 5000, 2002 ed.

dBA. Decibel, "A" scale. NFPA 1925, 2004 ed.

Dead Load. The weight of the aerial device structure and all materials, components, mechanisms, or equipment permanently fastened thereto. NFPA 1901, 2003 ed.

Debilitating Illness or Injury. A condition that temporarily or permanently prevents a member of the fire department from engaging in normal duties and activities as a result of illness or injury. NFPA 1500, 2002 ed.

Deck. A platform (floor) extending horizontally from one side of a ship to the other. NFPA 1405, 2001 ed.

Decontamination. The use of physical or chemical means to remove, inactivate, or destroy bloodborne, airborne, or foodborne pathogens on a surface or item to the point where they are no longer capable of transmitting infectious particles and the surface or item is rendered safe for handling, use, or disposal. NFPA 1581, 2005 ed.

Decontamination (Contamination Reduction). The physical and/or chemical process of reducing and preventing the spread of contamination from persons and equipment used at a hazardous materials incident. (See Figure D.1.) NFPA 471, 2002 ed.

Decontamination Corridor. The area usually located within the warm zone where decontamination procedures take place. NFPA 472, 2002 ed.

Dedicated Smoke Control System. A system that is intended for the purpose of smoke control only. It includes separate systems of air moving and distribution equipment that do not function under normal building operating conditions. NFPA 1, 2003 ed.

Deductive Reasoning. The process by which conclusions are drawn by logical inference from given premises. NFPA 921, 2004 ed.

Defect. A discontinuity in a part or a failure to function that interferes with the service or reliability for which the part was intended. NFPA 1901, 2003 ed.

Defective. Having a defect, or faulty. NFPA 1915, 2000 ed.

Figure D.1 Decontamination.

Defensible Space. An area as defined by the AHJ [typically a width of 9.14 m (30 ft) or more] between an improved property and a potential wildland fire where combustible materials and vegetation have been removed or modified to reduce the potential for fire on improved property spreading to wildland fuels or to provide a safe working area for fire fighters protecting life and improved property from wildland fire. NFPA 1144, 2002 ed.

Defensive Fire Fighting. The mode of manual fire control in which the only fire suppression activities taken are limited to those required to keep a fire from extending from one area to another. (See Figure D.2.) NFPA 600, 2005 ed.

Defensive Operations. Actions that are intended to control a fire by limiting its spread to a defined area, avoiding the commitment of personnel and equipment to dangerous areas. NFPA 1500, 2002 ed.

Defibrillation. The delivery of an electrical shock to the heart intended to reverse abnormal electrical activity. NFPA 450, 2004 ed.

Deficiency. The application of a component is not within its designed limits or specifications. NFPA 1071, 2000 ed.

Deflagration. Propagation of a combustion zone at a velocity that is less than the speed of sound in the unreacted medium. NFPA 68, 2002 ed.

Deflection. The deviation from a straight course or fixed direction. NFPA 1914, 2002 ed.

Deformation. Abnormal wear, defects, cracks or fractures, warpage, and deviations from the original condition that would affect safe and correct operation. NFPA 1915, 2000 ed.

Degradation. (a) A chemical action involving the molecular breakdown of a protective clothing material or equipment due to contact with a chemical. (b) The molecular breakdown of the spilled or released material to render it less hazardous during control operations. NFPA 471, 2002 ed.

Degree. Recognition of completion of a prescribed program of study at the postsecondary level. NFPA 1000, 2000 ed.

Degree-Granting Entity. An accredited institution of postsecondary higher education that is authorized to award degrees. NFPA 1000, 2000 ed.

Demand Zone Level. An area used to define or limit the management of a risk situation. NFPA 1720, 2004 ed.

Demonstrate. To show by actual performance. NFPA 472, 2002 ed.

Demonstration. The act of showing a skill. NFPA 1403, 2002 ed.

Density. The unit rate of water application to an area or surface expressed in gpm/ft^2 $[(L/min)/m^2]$. NFPA 15, 2001 ed.

Departure. An aircraft taking off from an airport. NFPA 402, 2002 ed.

Deployment. The procedures by which resources are distributed throughout the service area. NFPA 450, 2004 ed.

Deputy. A fully qualified individual who, in the absence of a superior, could be delegated the authority to manage a functional operation or perform a specific task. NFPA 1561, 2005 ed.

Descending (Line). A means of safely traveling down a fixed line using a descent control device. NFPA 1670, 2004 ed.

Descent Control Device. An auxiliary equipment system component; a friction or mechanical device utilized with rope to control descent. NFPA 1983, 2001 ed.

Describe. To explain verbally or in writing using standard terms recognized in the hazardous materials response community. NFPA 472, 2002 ed.

Design Load. The load for which a given piece of equipment or manufactured system was engineered for under normal static conditions. NFPA 1983, 2001 ed.

Design Verification Tests. Tests of a ladder structure and components thereof that are performed by the ladder manufacturer to prove conformance to design requirements and which can potentially compromise the integrity of the tested ladder. NFPA 1931, 2004 ed.

Designated Length. The length marked on the ladder. NFPA 1931, 2004 ed.

Detached Storage. Storage in a separate building or in an outside area located away from all structures. NFPA 1, 2003 ed.

Figure D.2 Defensive Fire Fighting.

Detection. (1) Sensing the existence of a fire, especially by a detector from one or more products of the fire, such as smoke, heat, ionized particles, infrared radiation, and the like. (See Figure D.3.) (2) The act or process of discovering and locating a fire. NFPA 921, 2004 ed.

Detention and Correctional Occupancy. An occupancy used to house four or more persons under varied degrees of restraint or security where such occupants are mostly incapable of self-preservation because of security measures not under the occupants' control. NFPA 5000, 2002 ed.

Detonation. Propagation of a combustion zone at a velocity that is greater than the speed of sound in the unreacted medium. NFPA 68, 2002 ed.

Detrimental Event. An incident or circumstance that produces or threatens to produce undesirable consequences to persons, property, or the environment that might ultimately be measured in terms of economic or financial loss. NFPA 1250, 2004 ed.

Develop. Modification, expansion, or compilation of existing educational or informational materials or resources. NFPA 1035, 2005 ed.

Dewatering. The process of removing water from a vessel. NFPA 1405, 2001 ed.

Diagnosis. The act of determining the cause of a problem. NFPA 1071, 2000 ed.

Diameter (Rope). The length of a straight line through the center of a rope. NFPA 1983, 2001 ed.

Dielectric Test Plane. A helmet term for the plane that runs from the intersection of the test line and midsagittal plane in the front of the headform diagonally through the headform to the intersection of the reference plane and midsagittal plane in the rear of the headform. NFPA 1976, 2000 ed.

Diffusion Flame. A flame in which fuel and air mix or diffuse together at the region of combustion. NFPA 921, 2004 ed.

Digital Radio System. A radio system that uses a binary representation of audio from one radio to another. NFPA 1221, 2002 ed.

Diluent. A gas used to dilute or reduce the concentration of oxygen. NFPA 53, 2004 ed.

Direct Attack. Fire fighting operations involving the application of extinguishing agents directly onto the burning fuel. (See Figure D.4.) NFPA 1145, 2000 ed.

Direct Exterior Window. A window in a communications center that faces an area that is not part of the secure area assigned solely to the communications center or that is accessible to the public. NFPA 1221, 2002 ed.

Direct or On-Line Medical Control or Oversight. The clinical advice or instructions given directly to emergency medical services (EMS) personnel by specially trained medical professionals. NFPA 450, 2004 ed.

Direct-Fired Vaporizer. A vaporizer in which heat furnished by a flame is directly applied to some form of heat exchange surface in contact with the liquefied petroleum gas to be vaporized. *Note: This classification includes submerged-combustion vaporizers.* NFPA 58, 2004 ed.

Disaster/Emergency Management Program. A program that implements the mission, vision, and strategic goals and objectives as well as the management framework of the program and organization. NFPA 1600, 2004 ed.

Discharge Outlet Size. The nominal size of the first fire hose connection from the pump. NFPA 1906, 2001 ed.

Discharge Pressure. The water pressure on the discharge manifold of the fire pump at the point of gauge attachment. NFPA 1911, 2002 ed.

Figure D.3 Common Residential Smoke Alarm and Detector.

Figure D.4 Direct Attack.

Discontinuity. A change in the normal, physical structure of a material that can affect its serviceability. NFPA 1914, 2002 ed.

Disentanglement. The cutting of a vehicle and/or machinery away from trapped or injured victims. NFPA 1670, 2004 ed.

Disinfectant. An agent that destroys, neutralizes, or inhibits the growth of harmful biological agents. NFPA 1851, 2001 ed.

Disinfection. The process used to inactivate virtually all recognized pathogenic microorganisms but not necessarily all microbial forms, such as bacterial endospore. NFPA 1581, 2005 ed.

Dispatch. To send out emergency response resources promptly to an address or incident location for a specific purpose. NFPA 450, 2004 ed.

Dispatch Circuit. A circuit over which an alarm is retransmitted automatically or manually from the communications center to an emergency response facility (ERF). NFPA 1221, 2002 ed.

Dispatch or Call Processing Interval. The interval between the time the dispatch agency makes its first contact with the caller and the time response resources are activated. NFPA 450, 2004 ed.

Dispatch Time. The point of receipt of the emergency alarm at the public safety answering point to the point where sufficient information is known to the dispatcher and applicable units are notified of the emergency. NFPA 1710, 2004 ed.

Dispatching. A process by which an alarm received at the communications center is retransmitted to emergency response facilities (ERFs) or to emergency response units in the field. NFPA 1221, 2002 ed.

Display Fireworks. Large fireworks articles designed to produce visible or audible effects for entertainment purposes by combustion, deflagration, or detonation. NFPA 1124, 2003 ed.

Display Screen. A general term that includes display devices that provide text, video, and graphics capabilities. NFPA 1221, 2002 ed.

Display Site. The immediate area where a fireworks display is conducted, including the discharge site, the fallout area, and the required separation distance from mortars to spectator viewing areas, but not spectator viewing areas or vehicle parking areas. NFPA 1123, 2000 ed.

Distillery. A plant or that portion of a plant where liquids produced by fermentation are concentrated and where the concentrated products are also mixed, stored, or packaged. NFPA 30, 2003 ed.

Distributor. A business engaged in the sale or resale, or both of compressed gases or cryogenic fluids, or both. NFPA 55, 2005 ed.

Dive. An exposure to increased pressure whether underwater or in a hyperbaric chamber. NFPA 1670, 2004 ed.

Dive Operation. A situation requiring divers to complete an assigned task. NFPA 1670, 2004 ed.

Dive Profile. Plan for a dive, including the depth and duration of the dive, in order to determine the level of nitrogen in the bloodstream. NFPA 1006, 2003 ed.

Dive Tables. Format utilized by divers, based upon various accepted studies, that calculates nitrogen levels and converts them to tabular data for determining a safe dive profile. NFPA 1006, 2003 ed.

Dive Team. An organization of public safety divers and members in training. NFPA 1670, 2004 ed.

Divemaster. Dive professional demonstrating an advanced level of competency, charged with coordinating and leading divers. NFPA 1006, 2003 ed.

Diver. An individual using breathing apparatus that supplies compressed breathing gas at the ambient pressure. NFPA 1670, 2004 ed.

Diverter Valve. A valve that, when actuated, diverts hydraulic fluid from one function to another or from one hydraulic system to another; in aerial devices, it is one valve that diverts hydraulic fluid from the stabilizers when the aerial device is in use and vice versa. NFPA 1914, 2002 ed.

Division. A supervisory level established to divide an incident into geographic areas of operations. NFPA 1561, 2005 ed.

Documentation. Any data or information supplied by the manufacturer or contractor relative to the apparatus, including information on its operation, service, and maintenance. NFPA 1901, 2003 ed.

Dormitory. A building or a space in a building in which group sleeping accommodations are provided for more than 16 persons who are not members of the same family in one room, or a series of closely associated rooms, under joint occupancy and single management, with or without meals, but without individual cooking facilities. NFPA 101, 2003 ed.

DOT. U.S. Department of Transportation. NFPA 57, 2002 ed.

DOT Cylinder. A pressure vessel, constructed and tested in accordance with 49 CFR 178.37, that is used for the storage and transportation of air under pressure. NFPA 1901, 2003 ed.

Double Block and Bleed. The closure of a line, duct, or pipe by closing, locking, and tagging two valves in line and opening, locking, and tagging a drain or vent valve inline between the two closed valves. NFPA 1006, 2003 ed.

Double Bottom. A void or tank space between the outer hull of the vessel and the floor of the vessel. NFPA 1405, 2001 ed.

Double-Row Racks. Two single-row racks placed back-to-back having a combined width up to 3.7 m (12 ft) with aisles of at least 1.1 m (3.5 ft) on each side. NFPA 13, 2002 ed.

Draft. A pressure difference that causes gases or air to flow through a chimney, vent, flue, or fuel burning equipment. NFPA 54, 2002 ed.

Drafting. The act of acquiring water for fire pumps from a static water supply by creating a negative pressure on the vacuum side of the fire pump. NFPA 1405, 2001 ed.

Drain Time. The time that it takes for a specified percent (customarily 25 percent) of the total solution that is contained in the foam to revert to liquid and drain out of the bubble structure. NFPA 1150, 2004 ed.

Drift. A time-dependent movement away from an established position. NFPA 1914, 2002 ed.

Drill. An exercise involving a credible simulated emergency that requires personnel to perform emergency response operations for the purpose of evaluating the effectiveness of the training and education programs and the competence of personnel in performing required response duties and functions. NFPA 601, 2005 ed.

Drip. A flow of liquid that lacks sufficient quantity or pressure to form a continuous stream. NFPA 1914, 2002 ed.

Driver's Enhanced Vision System (DEVS). An enhanced vision and navigation system for guiding aircraft rescue and fire fighting vehicles at night and during certain low-visibility conditions. *Note: The DEVS is comprised of three systems: (1) Navigation, which displays the ARFF vehicle's position on a moving map display mounted in the cab; (2) Tracking, which provides two-way digital communication between the ARFF vehicle and the Emergency Command Center; and (3) Vision, which allows the ARFF vehicle operator to see in 0/0 visibility conditions.* NFPA 414, 2001 ed.

Driver/Operator. A person having satisfactorily completed the requirements of driver/operator as specified in NFPA 1002, Standard for Fire Apparatus Driver/Operator Professional Qualifications. NFPA 1521, 2002 ed.

Drop Down. The spread of fire by the dropping or falling of burning materials. NFPA 921, 2004 ed. *Note: The term drop down is synonymous with the term* fall down. NFPA 921, 2004 ed.

Drug. Any substance, chemical, over-the-counter medication, or prescribed medication that can affect the performance of the fire fighter. NFPA 1500, 2002 ed.

Dry Bulk Terminal. A terminal equipped to handle dry goods that are stored in tanks and holds on the vessel. NFPA 1405, 2001 ed.

Dry Chemical. A powder composed of very small particles, usually sodium bicarbonate-, potassium bicarbonate-, or ammonium phosphate-based with added particulate material supplemented by special treatment to provide resistance to packing, resistance to moisture absorption (caking), and the proper flow capabilities. NFPA 17, 2002 ed.

Dry Hydrant. An arrangement of pipe permanently connected to a water source other than a piped, pressurized water supply system that provides a ready means of water supply for fire fighting purposes and that utilizes the drafting (suction) capability of fire department pumpers. (See Figure D.5.) NFPA 1144, 2002 ed.

Dry Location. A location not normally exposed to moisture such as in the interior of the driving or crew compartment, the interior of a fully enclosed walk-in fire apparatus body, or a watertight compartment opened only for maintenance operations. NFPA 1901, 2003 ed.

Dry Nitrogen. Nitrogen that has a dew point of −51°C (−60°F) or lower. NFPA 414, 2001 ed.

Dry Powder. Solid materials in powder or granular form designed to extinguish Class D combustible metal fires by crusting, smothering, or heat-transferring means. (See Figure D.6.) NFPA 402, 2002 ed.

Figure D.5 Dry-Barrel Hydrant.

FIGURE D.6 Extinguishing Fire with Dry Powder.

Dual-Use Industrial Chemicals. Highly toxic industrial chemicals that have been identified as mass casualty threats that could be used as weapons of terrorism to inflict casualties, generally on a civilian population, during a terrorist attack. *Note: Dual-use industrial chemicals can be liquid, solid, or gas agents.* NFPA 1994, 2001 ed.

Due Process. The compliance with the criminal and civil laws and procedures within the jurisdiction where the incident occurred. NFPA 1033, 2003 ed.

Dump Valve. A large opening from the water tank of a mobile water supply apparatus for unloading purposes. (SEE FIGURE D.7.) NFPA 1901, 2003 ed.

Dunnage. Loose packing material (usually wood) protecting a ship's cargo from damage or movement during transport. NFPA 1405, 2001 ed.

Duty. A major subdivision of the work performed by one individual. NFPA 1041, 2002 ed.

Duty Rating. The maximum load the ladder is designed to support when it is in use and properly positioned. NFPA 1931, 2004 ed.

Dynamic Balance. A physical condition that exists when a vehicle is driven into a turn under high speed and the vehicle displays no tendencies to pitch weight forward on the front steering wheels and exhibits no under steer or over steer conditions that could make the vehicle unstable. NFPA 414, 2001 ed.

Dynamic Suction Lift. The sum of the vertical lift and the friction and entrance loss caused by the flow through the suction strainers, sea chest, and piping, expressed in feet (meters). NFPA 1925, 2004 ed.

FIGURE D.7 Dump Valve.

Ear Covers. The integral part of the helmet designed to extend over and provide limited protection for the ears that does not provide significant thermal protection. NFPA 1976, 2000 ed.

Ease. The size requirements and tolerance of garments that allow good fit and do not inhibit the natural body movements or the performance of job-related tasks. NFPA 1977, 2005 ed.

Edge Protection. A means of protecting software components within a rope rescue system from the potentially harmful effects of exposed sharp or abrasive edges. NFPA 1670, 2004 ed.

Educational Methodology. The sum of knowledge and skills, including instructional materials, used by the public fire and life safety educator to create a positive outcome related to the learning objectives. NFPA 1035, 2005 ed.

Educational Occupancy. An occupancy used for educational purposes through the twelfth grade by six or more persons for 4 or more hours per day or more than 12 hours per week. NFPA 5000, 2002 ed.

Eductor. A device that uses the Venturi principle to siphon a liquid in a water stream. *Note: The pressure at the throat is below atmospheric pressure, allowing liquid at atmospheric pressure to flow into the water stream.* NFPA 1925, 2004 ed.

Effective Fire Temperatures. Temperatures reached in fires that identify physical effects that can be defined by specific temperature ranges. NFPA 921, 2004 ed.

Effective Operation. The accomplishment of or ability to accomplish the intended task. NFPA 1410, 2005 ed.

Effective Stream. A fire stream that has achieved and sustained the proper flow. NFPA 1410, 2005 ed.

Elasticity. The ability of an ensemble or element, when repeatedly stretched, to return to its original form as applied to wristlets and hoods. NFPA 1851, 2001 ed.

Electric Siren (Electromechanical). An audible warning device that produces sound by the use of an electric motor with an attached rotating slotted or perforated disc. NFPA 1901, 2003 ed.

Electric Spark. A small, incandescent particle created by some arcs. NFPA 921, 2004 ed.

Electrolyte. A liquid in which an electric current is easily conducted, for example, salt or brackish water. NFPA 1925, 2004 ed.

Electromagnetic Propagation. The travel of electromagnetic waves through a medium. NFPA 1221, 2002 ed.

Electronic Siren. An audible warning device that produces sound electronically through the use of amplifiers and electromagnetic speakers. NFPA 1901, 2003 ed.

Elements. The parts or items that comprise the protective ensemble, including coats, trousers, coveralls, helmets, gloves, footwear, and interface components. NFPA 1971, 2000 ed.

Elevating Platform. A self-supporting, turntable-mounted device consisting of a personnel-carrying platform attached to the uppermost boom of a series of power-operated booms that articulate, telescope, or both. NFPA 1912, 2001 ed.

Elevation Cylinder. The hydraulic components consisting of a cylinder barrel, cylinder rod, and related hardware that are used to vary the angle of the ladder or booms. NFPA 1914, 2002 ed.

Elevation Indicator. An instrument on an aerial device that shows the angle of elevation of the aerial ladder or boom. NFPA 1914, 2002 ed.

Elevation Lock. A manual- or positive-locking device (i.e., holding valve) that can be actuated to maintain indefinitely a desired angle or elevation without dependence upon engine power. NFPA 1914, 2002 ed.

Elongation. The increase in length, expressed in a percent of the original gauge length, that occurs in a sample of new rope when tested as specified herein. NFPA 1983, 2001 ed.

Emblems. Shields, heraldry, or printing that designates a governmental entity or a specific organization; rank, title, position, or other professional status that is painted, screened, embroidered, sewn, glued, bonded, or otherwise attached in a permanent manner to station/work uniform garments. NFPA 1975, 2004 ed.

Embrittlement. The hardening of a textile material that makes the ensemble or element or a textile material susceptible to easy fracture. NFPA 1851, 2001 ed.

Emergency. A fire, explosion, or hazardous condition that poses an immediate threat to the safety of life or damage to property. NFPA 1, 2003 ed.

Emergency Action Plan. A plan of designated actions by employers, employees, and other building occupants to ensure their safety from fire and other emergencies. NFPA 1620, 2003 ed.

Emergency Care First Responder (ECFR). An individual who has successfully completed the specified Emergency Care First Responder course developed by U.S. Department of Transportation and who holds an ECFR certification from the authority having jurisdiction. NFPA 473, 2002 ed.

Emergency Decontamination. The physical process of immediately reducing contamination of individuals in potentially life-threatening situations with or without the formal establishment of a decontamination corridor. NFPA 472, 1997 ed.

Emergency Hand-Crank Control. An auxiliary or supplemental control with which the operator can manually operate select functions of the aerial device. NFPA 1914, 2002 ed.

Emergency Incident. Any situation to which the emergency services organization responds to deliver emergency services, including rescue, fire suppression, emergency medical care, special operations, law enforcement, and other forms of hazard control and mitigation. NFPA 1561, 2005 ed.

Emergency Medical Care. The provision of treatment to patients, including first aid, cardiopulmonary resuscitation, basic life support (First Responder or EMT level), advanced life support (Paramedic level), and other medical procedures that occur prior to arrival at a hospital or other health care facility. (See Figure E.1.) NFPA 1710, 2004 ed.

Emergency Medical Dispatch. The receipt and management of requests for emergency medical assistance in the emergency medical services (EMS) system. NFPA 450, 2004 ed.

Emergency Medical Dispatcher (EMD). EMS personnel specifically trained and certified in interviewing techniques, pre-arrival instructions, and call prioritization. NFPA 450, 2004 ed.

Emergency Medical Examination Glove. An item of emergency medical protective clothing that is designed and configured to provide barrier protection to the wearer's hand to at least the wrist. NFPA 1999, 2003 ed.

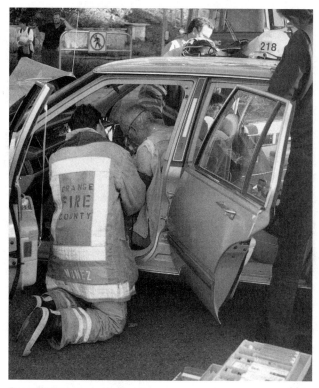

FIGURE E.1 Emergency Medical Care.

Emergency Medical Face Protection Device. An item of emergency medical protective clothing that is designed and configured to provide barrier protection to the wearer's face or head. NFPA 1999, 2003 ed.

Emergency Medical Footwear. An item of emergency medical protective clothing that is designed and configured to provide barrier protection to the wearer's feet. NFPA 1999, 2003 ed.

Emergency Medical Footwear Cover. An item of emergency medical protective clothing designed and configured to be worn over standard footwear to provide barrier and physical protection to the wearer's feet. NFPA 1999, 2003 ed.

Emergency Medical Garment. An item of emergency medical protective clothing designed and configured as a single garment or an assembly of multiple garments to provide barrier protection to the wearer's upper and lower torso, excluding the hands, face, and feet. NFPA 1999, 2003 ed.

Emergency Medical Operation. Delivery of emergency medical care and transportation prior to arrival at a hospital or other health care facility. (See Figure E.2.) NFPA 1581, 2005 ed.

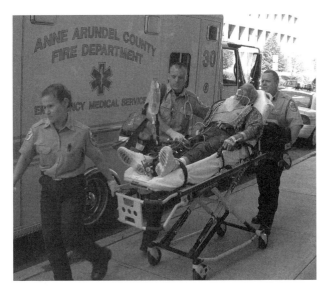

Figure E.2 Emergency Medical Operation.

Emergency Medical Protective Clothing. Multiple items of protective clothing, including garments, examination gloves, work gloves, cleaning gloves, footwear and footwear covers, and face protection devices designed and configured to provide limited physical protection and barrier protection against blood and body fluid-borne pathogens contact with the wearer's body during delivery of emergency patient care and other emergency medical functions. NFPA 1999, 2003 ed.

Emergency Medical Service (EMS). The organization(s) responsible for the care and transport of sick and injured persons to an appropriate emergency care facility. *Note: Referred to as Emergency Services in U.S. federal confined space regulations.* NFPA 1670, 2004 ed.

Emergency Medical Services. The provision of treatment such as first aid, cardiopulmonary resuscitation, basic life support, advanced life support, and other prehospital procedures including ambulance transportation to patients. NFPA 1500, 2002 ed.

Emergency Medical Services for Children (EMS-C). A national initiative to reduce child and youth disability and death from severe illness or injury. NFPA 450, 2004 ed.

Emergency Medical Technician (EMT). A person trained to administer emergency medical treatment more advanced than basic first aid. NFPA 424, 2002 ed.

Emergency Medical Technician-Ambulance (EMT-A). An individual who has completed a specified EMT-A course developed by the U.S. Department of Transportation and who holds an EMT-A certification from the authority having jurisdiction. NFPA 473, 2002 ed.

Emergency Medical Technician-Basic (EMT-B). A prehospital basic life support (BLS) provider with training based on the National Highway Traffic Safety Administration (NHTSA) National Standard Curriculum. NFPA 450, 2004 ed.

Emergency Medical Technician-Intermediate (EMT-I). An individual who has completed a course of instruction that includes selected modules of the U.S. Department of Transportation National Standard EMT-Paramedic Curriculum and who holds an intermediate level EMT-I or EMT-C certification from the authority having jurisdiction. NFPA 473, 2002 ed.

Emergency Medical Technician-Paramedic (EMT-P). An individual who has successfully completed a course of instruction that meets or exceeds the requirements of the U.S. Department of Transportation National Standard EMT-Paramedic Curriculum and who holds an EMT-P certification from the authority having jurisdiction. NFPA 473, 2002 ed.

Emergency Medical Work Glove. An item of emergency medical protective clothing that is designed and configured to provide physical and barrier protection to the wearer's hand and wrist. NFPA 1999, 2003 ed.

Emergency Operations. Activities of the fire department relating to rescue, fire suppression, emergency medical care, and special operations, including response to the scene of the incident and all functions performed at the scene. NFPA 1500, 2002 ed.

Emergency Operations Center. A fixed, designated area to be used in supporting and coordinating operations during emergencies. (See Figure E.3.) NFPA 424, 2002 ed.

Figure E.3 Emergency Operations Center.

Emergency Patient Care. The provision of treatment to patients, including first aid, cardiopulmonary resuscitation, basic life support, advanced life support, and other medical procedures that occur prior to arrival at a hospital or other health care facility. (See Figure E.4.) NFPA 1999, 2003 ed.

Emergency Relief Venting. An opening, construction method, or device that will automatically relieve excessive internal pressure due to an exposure fire. NFPA 30, 2003 ed.

Emergency Response Agency (ERA). An organization that engages in law enforcement, rescue, fire suppression, emergency medical, disaster assistance, and other related operations. NFPA 1221, 2002 ed.

Emergency Response Facility (ERF). A structure or a portion of a structure that houses emergency response agency equipment or personnel for response to alarms. NFPA 1221, 2002 ed.

Emergency Response Guidebook (NAERG). A reference book, written in plain language, to guide emergency responders in their initial actions at the incident scene. NFPA 472, 2002 ed.

Emergency Response Personnel (ERP). Personnel who respond to fire, emergency medical, hazardous materials, and other emergency situations for the preservation of life and property. NFPA 1221, 2002 ed.

Emergency Response Plan. A plan developed by an agency, with the cooperation of all participating agencies, that details specific actions to be performed by all personnel who are expected to respond during an emergency. NFPA 472, 2002 ed.

Emergency Response Vehicle. Any motorized vehicle designated by an organization or agency to respond to emergency incidents where provisions have been made to include warning systems and specialized components such as pumps, aerial devices, and rescue equipment and that is capable of transporting emergency response personnel. NFPA 1071, 2000 ed.

Emergency Scene. The area encompassed by the incident and the surrounding area needed by the emergency forces to stage apparatus and mitigate the incident. NFPA 901, 2001 ed.

Emergency Service System. A method of providing services through a planned and organized network of physical and human resources utilizing mandates with a defined mission. NFPA 1201, 2004 ed.

Emergency Services Organization (ESO). Any public, private, governmental, or military organization that provides emergency response and other related activities, whether for profit or government owned and operated. NFPA 1561, 2005 ed.

Emergency Vehicle Technician (EVT). An individual who performs inspection, diagnosis, maintenance, repair, and testing activities on emergency response vehicles and who, by possession of a recognized certificate, professional standing, or skill, has acquired the knowledge, training, and experience and has demonstrated the ability to deal with issues related to the subject matter, the work, or the project. NFPA 1071, 2000 ed.

Empennage. The tail assembly of an aircraft, which includes the horizontal and vertical stabilizers. NFPA 402, 2002 ed.

Employee Illness and Injury. A work-related illness or injury requiring evaluation or medical follow-up. NFPA 450, 2004 ed.

EMS. Emergency medical services. NFPA 1925, 2004 ed.

EMS System. A comprehensive, coordinated arrangement of resources and functions that are organized to respond in a timely, staged manner to medical emergencies regardless of their cause. NFPA 450, 2004 ed.

EMS/HM Level I Responder. EMS personnel at EMS/HM Level I are those persons who, in the course of their normal duties, might be called on to perform patient care activities in the cold zone at a hazardous materials incident. *Note: EMS/HM Level I responders provide care only to those individuals who no longer pose a significant risk of secondary contamination.* NFPA 473, 2002 ed.

Figure E.4 Emergency Patient Care.

EMS/HM Level II Responder. Personnel at EMS/HM Level II are those persons who, in the course of their normal activities, might be called upon to perform patient care activities in the warm zone at hazardous materials incidents. *Note: EMS/HM Level II responder personnel might be required to provide care to those individuals who still pose a significant risk of secondary contamination. In addition, personnel at this level are able to coordinate EMS activities at a hazardous materials incident and provide medical support for hazardous materials response personnel.* NFPA 473, 2002 ed.

En Route Interval. A measurement that begins at the time a response unit starts to move toward an incident, and the time the unit comes to a complete stop at the location of the incident. NFPA 450, 2004 ed.

Encapsulating. A type of ensemble that provides vapor- or gastight protection, or liquidtight protection, or both, and completely covers the wearer and the wearer's respirator. (See Figure E.5.) NFPA 1992, 2005 ed.

Enclosed Compartment. An area designed to protect stored items from environmental damage (weather resistant) that is confined on six sides and equipped with an access opening(s) that can be closed and latched. NFPA 1901, 2003 ed.

Enclosed Parking Structure. Any parking structure that is not an open parking structure. NFPA 88A, 2002 ed.

Enclosed Structure. A structure with a roof or ceiling and at least two walls that can present fire hazards to employees such as accumulations of smoke, toxic gases, and heat, similar to those found in buildings. NFPA 600, 2005 ed.

End-of-Service-Time Indicator. A warning device on an SCBA that warns the user that the end of the service time of the SCBA is approaching. NFPA 1981, 2002 ed.

Endangered Area. The actual or potential area of exposure from a hazardous material. NFPA 472, 2002 ed.

Energy Absorbing System. A material, suspension system, or combination thereof incorporated into the design of the helmet to attenuate impact energy. NFPA 1971, 2000 ed.

Engine. A fire department pumper that has a rated capacity of 2840 L/min (750 gpm) or more. (See Figure E.6.) NFPA 1410, 2005 ed.

Engine Company. A group of fire fighters who work as a unit and are equipped with one or more pumping engines that have rated capacities of 2840 L/min (750 gpm) or more. NFPA 1410, 2005 ed.

Engine-Driven Generator. A generator driven by an internal combustion engine. NFPA 1221, 2002 ed.

Engineering Controls. Physical features or mechanical processes within fixed facilities or vehicles that are implemented to improve efficiency, safety, or comfort associated with their operation or use. NFPA 1581, 2005 ed.

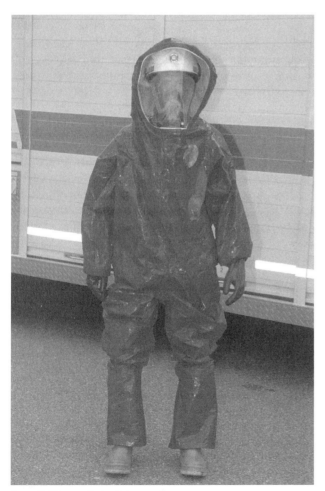

FIGURE E.5 Encapsulating Suit.

FIGURE E.6 Engine Company Apparatus.

Engulfment. The surrounding and effective capture of a person by a fluid (e.g., liquid, finely divided particulate) substance that can be aspirated to cause death by filling or plugging the respiratory system or that can exert enough force on the body to cause death by strangulation, constriction, or crushing. NFPA 1670, 2004 ed.

Enhanced 9-1-1. Emergency telephone service that provides selective routing, automatic number identification (ANI), and automatic location identification (ALI). (See Figure E.7.) NFPA 1221, 2002 ed.

Ensemble. *As applied to fire fighting:* Multiple elements of clothing and equipment, including coats, trousers, coveralls, helmets, gloves, footwear, and interface components, designed to provide a degree of protection for fire fighters from adverse exposures to the inherent risks of structural fire fighting operations and certain other emergency operations. NFPA 1851, 2001 ed. *As applied to chemical/biological terrorism: Ensemble* is a shortened term for Chemical/Biological Terrorism Incident Protective Ensemble. NFPA 1994, 2001 ed.

Ensemble Elements. The parts or items that comprise the chemical/biological terrorism incident protective ensemble. NFPA 1994, 2001 ed.

Entity. A governmental agency or jurisdiction, private or public company, partnership, nonprofit organization, or other organization that has disaster/emergency management and continuity of operations responsibilities. NFPA 1600, 2004 ed.

Entrainment. The process of air or gases being drawn into a fire, plume, or jet. NFPA 921, 2004 ed.

Entry. The action by which a person passes into a confined space. *Note: Entry includes ensuing work or rescue activities in that environment and is considered to have occurred as soon as any part of the entrant's body breaks the plane of an opening into the space, trench, or excavation.* NFPA 1670, 2004 ed.

Entry Fire Fighting. Extraordinarily specialized fire fighting operations that can include the activities of rescue, fire suppression, and property conservation at incidents involving fires producing very high levels of conductive, convective, and radiant heat, such as aircraft fires, bulk flammable gas fires, and bulk flammable liquid fires. NFPA 1500, 2002 ed.

Entry Permit. A written or printed document, established by an employer, for nonrescue entry into confined spaces. NFPA 1670, 2004 ed.

Entry Team. The group of individuals, with established communications and leadership, assigned to perform work or rescue activities beyond the opening of, and within, the space, trench, or excavation. NFPA 1670, 2004 ed.

Environment. A collection of characteristics such as weather, altitude, and terrain contained in an area that are unique to a location. NFPA 1670, 2004 ed.

Environmental Hazard. A condition capable of posing an unreasonable risk to air, water, or soil quality and to plants or wildlife. NFPA 471, 2002 ed.

Environmental Surface. Interior patient care areas, both stationary and in vehicles, and other surfaces not designed for intrusive contact with the patient or contact with mucosal tissue. NFPA 1581, 2005 ed.

Equipment Allowance. Any equipment added to the vehicle that is not directly required for the vehicle to discharge water or other fire fighting agent(s) on the initial attack. NFPA 414, 2001 ed.

Escape. Immediate self-rescue of a single fire or emergency services person from a life-threatening emergency situation, generally above ground, using system components or manufactured systems designed for self-rescue escape. NFPA 1983, 2001 ed.

Escape Belt. A belt that is certified as compliant with the applicable requirements of NFPA 1983 for an escape belt, and that is intended for use only by the wearer as an emergency self-rescue device. NFPA 1983, 2001 ed.

Escape Descent Control Device. An auxiliary equipment system component; a friction or mechanical device utilized with escape rope to control descent. NFPA 1983, 2001 ed.

Escape Rope. A system component; a single-purpose, one-time use, emergency self-escape (self-rescue) rope; not classified as a life safety rope. (See Figure E.8.) NFPA 1983, 2001 ed.

Escape Trunk. A vertical trunk fitted with a ladder to allow personnel to escape if trapped. NFPA 1405, 2001 ed.

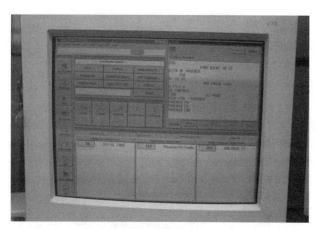

Figure E.7 Automatic Location Identification.

FIGURE E.8 Personal Escape Rope.

Essential Job Task. Task or assigned duty that is critical to successful performance of the job. NFPA 1582, 2003 ed.

Evacuation. The temporary movement of people and their possessions from locations threatened by wildland fire. NFPA 1144, 2002 ed.

Evacuation Capability. The ability of occupants, residents, and staff as a group either to evacuate a building or to relocate from the point of occupancy to a point of safety. NFPA 101, 2003 ed.

Evacuation Plan. A plan specifying safe and effective methods for the temporary movement of people and their possessions from locations threatened by wildland fire. NFPA 1051, 2002 ed.

Evacuation Time. The elapsed time between an aircraft accident/incident and the removal of all surviving occupants. NFPA 402, 2002 ed.

Evacuee. An aircraft occupant who has exited the aircraft following an accident/incident. NFPA 402, 2002 ed.

Evolution. A set of prescribed actions that result in an effective fireground activity. NFPA 1410, 2005 ed.

Examination Glove. *Examination glove* is a shortened term for emergency medical examination glove. NFPA 1999, 2003 ed.

Excavation. Any man-made cut, cavity, trench, or depression in an earth surface, formed by the removal of earth. NFPA 1670, 2004 ed.

Excepted Packaging. Range from product's fiberboard box to a sturdy wooden or steel crate, and may include limited quantities of materials, instruments, and articles such as smoke detectors. NFPA 472, 2002 ed.

Excess Flow Control. A fail-safe system or approved means designed to shut off flow due to a rupture in pressurized piping systems. NFPA 55, 2005 ed.

Excess Flow Valve. A valve inserted into a compressed gas cylinder, portable tank, or stationary tank that is designed to positively shut off the flow of gas in the event that its predetermined flow is exceeded. NFPA 1, 2003 ed.

Exhalation Valve. A device that allows exhaled air to leave a facepiece and prevents outside air from entering through the valve. NFPA 1404, 2002 ed.

Exhaust Valve. One-way vent that releases exhaust to the outside environment and prevents entry of outside environment. NFPA 1991, 2005 ed.

Existing. That which is already in existence on the date an edition of a code or standard goes into effect. NFPA 101, 2003 ed.

Existing Barrier. See Existing Building.

Existing Building. A building erected or officially authorized prior to the effective date of the adoption of an edition of a Code by the agency or jurisdiction. NFPA 101, 2003 ed.

Existing Condition. Any situation, circumstance, or physical makeup of any structure, premise, or process that was ongoing or in effect prior to the adoption of a standard. NFPA 1141, 2003 ed.

Exit. That portion of a means of egress that is separated from all other spaces of a building or structure by construction or equipment as required to provide a protected way of travel to the exit discharge. NFPA 101, 2003 ed.

Exit Access. That portion of a means of egress that leads to an exit. NFPA 101, 2003 ed.

Exit Discharge. That portion of a means of egress between the termination of an exit and a public way. NFPA 101, 2003 ed.

Expansion Ratio. The ratio of the volume of the foam in its aerated state to the original volume of the non-aerated foam solution. NFPA 1901, 2003 ed.

Explosion. The sudden conversion of potential energy (chemical, mechanical, or nuclear) into kinetic energy that produces and violently releases gas. NFPA 69, 2002 ed.

Explosive. Any chemical compound, mixture, or device, the primary or common purpose of which is to function by explosion. (SEE FIGURE E.9.) NFPA 495, 2001 ed.

Explosive Material. Any explosive, blasting agent, emulsion explosive, water gel, or detonator. NFPA 295, 2001 ed.

Exposed Surface. The side of a structural assembly or object that is directly exposed to the fire. NFPA 921, 2004 ed.

Exposure. The process by which people, animals, the environment, and equipment are subjected to or come in contact with a hazardous material. NFPA 472, 2002 ed.

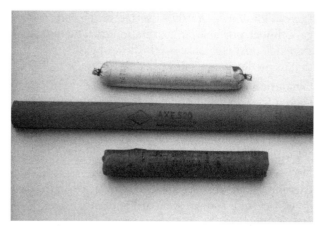

Figure E.9 Explosives.

Figure E.10 Extension Ladder.

Exposure. The state of being exposed to loss because of some hazard or contingency. NFPA 1250, 2004 ed.

Exposure Hazard. A structure within 50 ft (15.24 m) of another building and 100 ft² (9.3 m²) or larger in area. NFPA 1142, 2001 ed.

Exposure Incident. A specific eye, mouth, other mucous membrane, nonintact skin, or parenteral contact with blood, body fluids, or other potentially infectious materials, or inhalation of airborne pathogens, ingestion of foodborne pathogens and/or toxins. NFPA 1582, 2003 ed.

Extendable Turret. A device, permanently mounted with a power-operated boom or booms, designed to supply a large-capacity, mobile, elevatable water stream or other fire-extinguishing agents, or both. NFPA 414, 2001 ed.

Extended Attack. A wildland fire that cannot be controlled by initial attack resources within an established period of time as determined by the AHJ and additional fire fighting resources are arriving, en route, or being ordered by the incident commander. NFPA 1051, 2002 ed.

Extension Cylinder. The hydraulic components consisting of a cylinder barrel, cylinder rod, and related hardware that are used to vary the length of extension of a telescoping aerial device. NFPA 1914, 2002 ed.

Extension Indicator. A device on an aerial ladder or extensible boom aerial device that indicates the number of feet that the device has been extended. NFPA 1914, 2002 ed.

Extension Ladder. A non-self-supporting ground ladder that consists of two or more sections traveling in guides, brackets, or the equivalent arranged so as to allow length adjustment. (See Figure E.10.) NFPA 1931, 2004 ed.

Extension Sheave. A pulley through which an extension cable operates. NFPA 1914, 2002 ed.

Exterior. A nonsheltered location exposed to the environment, either continuously or intermittently. NFPA 1901, 2003 ed.

External Fitting. Any fitting externally located on, and part of, the ensemble that is not part of the garment material, visor material, gloves, footwear, seams, or closure assembly that allows the passage of gases, liquids, or electrical current from the outside to the inside of the element or item. NFPA 1991, 2005 ed.

Extinguish. To cause to cease burning. NFPA 921, 2004 ed.

Extinguishing Agent Compatibility. Related to the requirement that the chemical composition of each agent be such that one will not adversely affect the performance of other agents that might be used on a common fire. NFPA 402, 2002 ed.

Extinguishing Device. A device used to suppress fire, including, but not limited to, CO_2 extinguishers, dry chemical extinguishers, hose lines, and fire fighting foam. (See Figure E.11.) NFPA 1006, 2003 ed.

Extra (High) Hazard. An occupancy in which the total amount of Class A combustibles and Class B flammables present, in storage, production, use, finished product, or combination thereof, is over and above those expected in occupancies classed as ordinary (moderate) hazard. *Note: Extra (high) hazard occupancies could consist of woodworking, vehicle repair, aircraft and boat servicing, cooking areas, individual product display showrooms, product convention center displays, and storage and manufacturing processes such as painting, dipping, and coating, including flammable liquid handling. Also included is warehousing of or in-process storage of other than Class I and Class II commodities as defined by NFPA 13, Standard for the Installation of Sprinkler Systems.* (See Figure E.12.) NFPA 1, 2003 ed.

FIGURE E.11 Examples of Extinguishing Devices.

Extrication. The removal of trapped victims from a vehicle or machinery. NFPA 1670, 2004 ed.

Eye and Face Protection Device. An element of the protective ensemble intended to protect the wearer's eyes and face. NFPA 1951, 2005 ed.

Eye/Face Positioning Index. The vertical distance, as specified by the helmet manufacturer, from the top lateral midpoint of the faceshield components to the basic plane of the Alderson 50th percentile adult male headform where the faceshield component is positioned on the headform. NFPA 1971, 2000 ed.

Eyelet. A reinforced hole placed in the footwear upper throat through which laces are threaded to secure the footwear to the wearer. NFPA 1977, 2005 ed.

Eyerow. The row of eyelets. NFPA 1977, 2005 ed.

FIGURE E.12 Example of Extra (High) Hazard Area.

Fabric Component. Any single or combination of natural or synthetic material(s) that are pliable and that are made by weaving, felting, forming, or knitting. NFPA 1981, 2002 ed.

Face Gasket. The water pressure seal at the mating surfaces of nonthreaded couplings or adapters. NFPA 1963, 1998 ed.

Face Protection Device. A device constructed of protective materials, designed and configured to cover part or all of the wearer's face or head. NFPA 1581, 2005 ed. *Note:* Face protection device *is also a shortened term for* emergency medical face protection device.

Face(s) (also Wall, Side, or Belly). The vertical or inclined earth surface formed as a result of excavation work. NFPA 1006, 2003 ed.

Facepiece. The component of a respirator that covers the wearer's nose, mouth, and eyes. NFPA 1404, 2002 ed.

Faceshield. A helmet component not intended as primary eye protection, but to help protect a portion of the wearer's face in addition to the eyes. NFPA 1971, 2000 ed.

Facility. *As applied to industrial fire brigade:* A structure or building located on a site that serves a particular purpose. NFPA 1081, 2001 ed. *As applied to rocket motors:* All land and buildings, including the rocket motor plant, constituting a model or high power rocket motor manufacturing operation. NFPA 1125, 2001 ed.

Facility Emergency Action Plan (Detention and Correctional Occupancy). A plan developed by a detention and correctional facility to describe the actions to be taken during various internal and external emergencies. The plan is generally available on site. NFPA 1620, 2003 ed.

Facility Emergency Action Plan (Health Care Occupancy). A plan developed by a health care facility to describe the actions to be taken during various internal and external emergencies. The plan is generally available on site. NFPA 1620, 2003 ed.

FAD. Free air delivery. NFPA 1901, 1999 ed.

Failure. *As applied to search and rescue:* The breakage, displacement, or permanent deformation of a structural member or connection so as to reduce its structural integrity and its supportive capabilities. NFPA 1670, 2004 ed. *As applied to fire apparatus:* A cessation of proper functioning or performance. NFPA 1915, 2000 ed.

Failure Analysis. A logical, systematic examination of an item, component, assembly, or structure and its place and function within a system, conducted in order to identify and analyze the probability, causes, and consequences of potential and real failures. NFPA 921, 2004 ed.

Fall Factor. A measure of fall severity calculated by dividing the distance fallen by the length of rope used to arrest the fall. NFPA 1983, 2001 ed.

Family Dynamics. The structure and characteristics of a person's living environment(s), including relatives, caregivers, other relationships, and their interactions with each other. NFPA 1035, 2005 ed.

Fantail. The stern overhang of a ship. NFPA 1405, 2001 ed.

Farms. Those properties that are used primarily for agricultural purposes. NFPA 1141, 2003 ed.

Fastener. A mechanical device, such as a rivet, bolt, screw, or pin, that is used to hold two or more components together securely. NFPA 1914, 2002 ed.

Fatality. An injury that is fatal or becomes fatal within one year of the incident. NFPA 901, 2001 ed.

Federal Aviation Administration (FAA). An agency of the U. S. federal government charged with the primary responsibility of regulating aviation activities. NFPA 402, 2002 ed.

Federal Motor Vehicle Safety Standard (FMVSS). Regulations promulgated by National Highway Traffic Safety Administration (NHTSA) of the United States under Public Law 89-563 that are mandatory and must be complied with when vehicles or items of motor vehicle equipment are manufactured and certified thereto. NFPA 1901, 2003 ed.

Federal Response Plan. A U.S. government plan for the basic mechanisms and structures by which the federal government will mobilize resources and conduct activities to augment state and local disaster and emergency response efforts. NFPA 1670, 2004 ed.

Feedback. Comments and opinions regarding the system to be reviewed for appropriate changes or modifications. NFPA 1401, 2001 ed.

FEMA Task Force Search and Rescue Marking System. Distinct markings made with international orange spray paint near a collapsed structure's most accessible point of entry. NFPA 1670, 2004 ed.

FEMA Task Force Structure Marking System, Structure Identification Within a Geographic Area. Distinct markings made with international orange spray paint to label buildings with their street number so that personnel can differentiate one building from another. NFPA 1670, 2004 ed.

FEMA Task Force Structure/Hazard Evaluation Marking System. Distinct markings made with international orange spray paint, after performing a building hazard identification, near a collapsed structure's most accessible point of entry. NFPA 1670, 2004 ed.

Ferromagnetic Materials. Materials that can be magnetized and strongly attracted to a magnetic field such as iron, steel, cobalt, and nickel. NFPA 1914, 2002 ed.

Field Test. The non-laboratory evaluation of one or more protective ensemble elements used to determine product performance related to organizational expectations or to compare products in a manner related to their intended use. NFPA 1851, 2001 ed.

Fill Hose. Flexible hose plumbed to connect SCBA cylinders to the compressed air supply for filling purposes. NFPA 1901, 1999 ed.

Fill Station. See SCBA Fill Station.

Film-Forming Fluoroprotein (FFFP) Foam. A protein-based foam concentrate incorporating fluorinated surfactants that form a foam capable of producing a vapor-suppressing, aqueous film on the surface of hydrocarbon fuels. (SEE FIGURE F.1.) Note: *FFFP might show an acceptable level of compatibility to dry chemicals and might be suitable for use with those agents.* NFPA 412, 2003 ed.

Finance. The incident management section responsible for all incident costs and financial considerations. NFPA 1143, 2003 ed.

Findings. All materials used in the construction of items, excluding textiles and interlinings. NFPA 1975, 2004 ed.

Finish Rating. The time in minutes, determined under specific laboratory conditions, at which the stud or joist in contact with the exposed protective membrane in a protected combustible assembly reaches an average temperature rise of 121°C (250°F) or an individual temperature rise of 163°C (325°F) as measured behind the protective membrane nearest the fire on the plane of the wood. NFPA 921, 2004 ed.

Fire. A rapid oxidation process, which is a chemical reaction resulting in the evolution of light and heat in varying intensities. NFPA 921, 2004 ed.

Fire Alarm Signal. A signal initiated by a fire alarm-initiating device such as a manual fire alarm box, automatic fire detector, waterflow switch, or other device in which activation is indicative of the presence of a fire or fire signature. (SEE FIGURE F.2.) NFPA 72, 2002 ed.

FIGURE F.1 Applying FFFP Foam.

FIGURE F.2 Fire Alarm Notification Device.

Fire Alarm System. A system or portion of a combination system that consists of components and circuits arranged to monitor and annunciate the status of fire alarm or supervisory signal-initiating devices and to initiate the appropriate response to those signals. (See Figure F.3.) NFPA 72, 2002 ed.

Fire Analysis. The process of determining the origin, cause, development, and responsibility as well as the failure analysis of a fire or explosion. NFPA 921, 2004 ed.

Fire and Emergency Services Personnel. Members of fire departments, other governmental agencies, or other organizations that have the public safety responsibilities and who would respond to terrorism incidents where a chemical terrorism agent(s), biological terrorism agent(s), or dual-use industrial chemical(s) has been or could be released. NFPA 1994, 2001 ed.

Fire Apparatus. A vehicle designed to be used under emergency conditions to transport personnel and equipment, and to support the suppression of fires and mitigation of other hazardous situations. NFPA 1901, 2003 ed.

Fire Apparatus Driver/Operator. A fire department member who is authorized by the authority having jurisdiction to drive, operate, or both drive and operate fire department vehicles. NFPA 1451, 2002 ed.

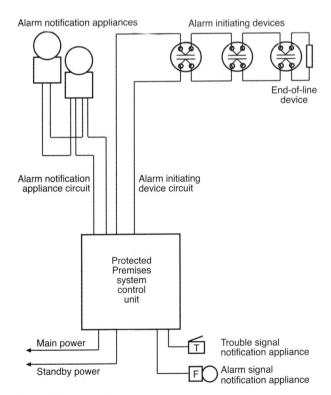

FIGURE F.3 Fire Alarm System.

Fire Area. *As applied to buildings:* An area of a building separated from the remainder of the building by construction having a fire resistance of at least 1 hour and having all communicating openings properly protected by an assembly having a fire resistance rating of at least 1 hour. NFPA 30, 2003 ed. *As applied to flammable and combustible liquids:* An area that is physically separated from other areas by space, barriers, walls, or other means in order to contain fire within that area. NFPA 5000, 2002 ed. *As applied to incident reporting:* The space within a structure bounded by fire division assemblies (2-hour fire rating or greater). NFPA 901, 2001 ed. *As applied to wildland fire fighting:* The area within wildland fire perimeter control lines. NFPA 901, 2001 ed.

Fire Barrier. A continuous membrane or a membrane with discontinuities created by protected openings with a specified fire protection rating, where such membrane is designed and constructed with a specified fire resistance rating to limit the spread of fire, that also restricts the movement of smoke. NFPA 101, 2003 ed.

Fire Behavior. The manner in which a fire reacts to the variables of fuels, weather, and topography. NFPA 1051, 2002 ed.

Fire Blackout. That point in time when there is no longer any evidence of open flame or glow of burned material. NFPA 901, 2001 ed.

Fire Brigade. A group of people organized to engage in rescue, fire suppression, and related activities. NFPA 1561, 2005 ed.

Fire Casualty. A person who is injured or dies at the scene of a fire, whether from natural causes, direct involvement with the fire, or an accident sustained while involved in fire control, a rescue attempt, or escaping from the dangers of the fire. NFPA 901, 2001 ed.

Fire Cause. The circumstances, conditions, or agencies that bring together a fuel, ignition source, and oxidizer (such as air or oxygen) resulting in a fire or a combustion explosion. NFPA 921, 2004 ed.

Fire Chief. The highest ranking officer in charge of a fire department. NFPA 1710, 2004 ed.

Fire Compartment. A space within a building that is enclosed by fire barriers on all sides, including the top and bottom. NFPA 101, 2003 ed.

Fire Contained. That point in time when fire spread is stopped but the fire is not necessarily under control. NFPA 901, 2001 ed.

Fire Control Line. Comprehensive term for all constructed or natural barriers and treated fire edges used to control a fire. NFPA 901, 2001 ed.

Fire Control Measures. Methods used to secure ignition sources at an incident scene that can include hoseline placement and utilization of chemical agents to suppress fire potential. NFPA 1006, 2003 ed.

Fire Control Plan. A set of general arrangement plans that illustrate, for each deck, the fire control stations, fire-resisting bulkheads, and fire-retarding bulkheads, together with particulars of the fire-detecting, manual alarm, and fire-extinguishing systems, fire doors, means of access to different compartments, and ventilating systems, including locations of dampers and fan controls. NFPA 1405, 2001 ed.

Fire Damage. The total damage to a building, structure, vehicle, natural vegetation cover, or outside property resulting from a fire and the act of controlling that fire. NFPA 901, 2001 ed.

Fire Department. An organization providing rescue, fire suppression, and related activities, including any public, governmental, private, industrial, or military organization engaging in this type of activity. (SEE FIGURE F.4.) NFPA 1002, 2003 ed.

Fire Department Access Road. The road or other means developed to allow access and operational setup for fire fighting and rescue apparatus. NFPA 1, 2003 ed.

Fire Department Facility. Any building or area owned, operated, occupied, or used by a fire department on a routine basis. NFPA 1500, 2002 ed.

Fire Department Ground Ladder. Any portable ladder specifically designed for fire department use in rescue, fire fighting operations, or training. (SEE FIGURE F.5.) NFPA 1931, 2004 ed.

Fire Department Physician. A licensed doctor of medicine or osteopathy who has been designated by the fire department to provide professional expertise in the areas of occupational safety and health as they relate to emergency services. NFPA 1582, 2003 ed.

Fire Department Pumper. A piece of fire apparatus with a permanently mounted fire pump that has a rated discharge capacity of 750 gpm (2850 L/min) or greater as defined in NFPA 1901. NFPA 1002, 2003 ed.

Fire Department Safety Officer. Functions comprised of the health and safety officer and the incident safety officer. These roles can be performed by one member or several members as designated by the fire chief. NFPA 1521, 2002 ed.

Fire Department Vehicle. Any vehicle, including fire apparatus, operated by a fire department. NFPA 1002, 2003 ed.

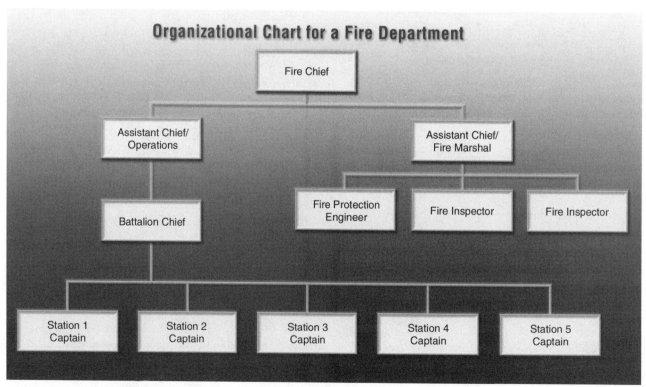

FIGURE F.4 Typical Fire Department Organization.

FIGURE F.5 Stored Portable Ladders.

Fire Division Assembly. A "fire-rated assembly" that has a fire resistance rating of 2 test hours or longer. NFPA 901, 2001 ed.

Fire Division Compartment. A complete compartment surrounded on all sides by fire-rated assemblies with a 2-hour fire protection rating or more. NFPA 901, 2001 ed.

Fire Door Assembly. Any combination of a fire door, a frame, hardware, and other accessories that together provide a specific degree of fire protection to the opening. NFPA 80, 1999 ed.

Fire Dynamics. The detailed study of how chemistry, fire science, and the engineering disciplines of fluid mechanics and heat transfer interact to influence fire behavior. NFPA 921, 2004 ed.

Fire Extinguished. That point in time when there is no longer any abnormal heat or smoke being generated in material that was previously burning. NFPA 901, 2001 ed.

Fire Fighter Candidate. The person who has fulfilled the entrance requirements but has not met the job performance requirements for Fire Fighter I. NFPA 1001, 2002 ed.

Fire Fighter I. The person, at the first level of progression, who has demonstrated the knowledge and skills to function as an integral member of a fire fighting team under direct supervision in hazardous conditions. NFPA 1001, 2002 ed.

Fire Fighter II. The person, at the second level of progression, who has demonstrated the skills and depth of knowledge to function under general supervision. NFPA 1001, 2002 ed.

Fire Fighting Vessel. Any vessel whose primary mission is fire fighting and pumping operations, including emergency operations. NFPA 1925, 2004 ed.

Fire Flow. The flow rate of a water supply, measured at 138 kPa (20 psi) residual pressure, that is available for fire fighting. NFPA 1141, 2003 ed.

Fire Growth Potential. The potential size or intensity of a fire over a period of time based on the available fuel and the fire's configuration. NFPA 1031, 2003 ed.

Fire Hose. A flexible conduit used to convey water. (See Figure F.6.) NFPA 1961, 2002 ed.

Fire Hose Appliance. A piece of hardware (excluding nozzles) generally intended for connection to fire hose to control or convey water. NFPA 1965, 2003 ed.

Fire Hydrant. A valved connection on a water supply system having one or more outlets and that is used to supply hose and fire department pumpers with water. (See Figure F.7.) NFPA 1141, 2003 ed.

Fire Inspector I. An individual at the first level of progression who has met the job performance requirements specified for Level I. *Note: The Fire Inspector I conducts basic fire inspections and applies codes and standards.* NFPA 1031, 2003 ed.

Fire Inspector II. An individual at the second or intermediate level of progression who has met the job performance requirements for Level II. *Note: The Fire Inspector II conducts most types of inspections and interprets applicable codes and standards.* NFPA 1031, 2003 ed.

Fire Inspector III. An individual at the third and most advanced level of progression who has met the job performance requirements specified for Level III. *Note: The Fire Inspector III performs all types of fire inspections, plans review duties, and resolves complex code-related issues.* NFPA 1031, 2003 ed.

Fire Investigation. The process of determining the origin, cause, and development of a fire or explosion. (See Figure F.8.) NFPA 921, 2004 ed.

FIGURE F.6 Fire Hose.

FIGURE F.7 Fire Hydrant.

FIGURE F.8 Fire Investigation.

Fire Investigator. An individual who has demonstrated the skills and knowledge necessary to conduct, coordinate, and complete an investigation. NFPA 1033, 2003 ed.

Fire Lane. The road or other means developed to allow access and operational setup for fire fighting and rescue apparatus. NFPA 1, 2000 ed.

Fire Officer I. The fire officer, at the supervisory level, who has met the job performance requirements for Level I. NFPA 1021, 2003 ed.

Fire Officer II. The fire officer, at the supervisory/managerial level, who has met the job performance requirements for Level II. NFPA 1021, 2003 ed.

Fire Officer III. The fire officer, at the managerial/administrative level, who has met the job performance requirements for Level III. NFPA 1021, 2003 ed.

Fire Officer IV. The fire officer, at the administrative level, who has met the job performance requirements specified for Level IV. NFPA 1021, 2003 ed.

Fire Patterns. The visible or measurable physical effects that remain after a fire. (See Figure F.9.) NFPA 921, 2004 ed.

Fire Point. The lowest temperature at which a liquid will ignite and achieve sustained burning when exposed to a test flame in accordance with ASTM D 92, Standard Test Method for Flash and Fire Points by Cleveland Open Cup. NFPA 30, 2003 ed.

Fire Protection. Methods of providing for fire control or fire extinguishment. NFPA 801, 2003 ed.

Fire Protection System. Any fire alarm device or system or fire extinguishing device or system, or their combination, that is designed and installed for detecting, controlling, or extinguishing a fire or otherwise alerting occupants, or the fire department, or both, that a fire has occurred. NFPA 1141, 2003 ed.

Fire Pump. A water pump with a rated capacity of 250 gpm (1000 L/min) or greater at 150 psi (10 bar) net pump pressure that is mounted on a fire apparatus and used for fire fighting. NFPA 1901, 2003 ed.

FIGURE F.9 V-Pattern.

Fire Resistance Rating. The time, in minutes or hours, that materials or assemblies have withstood a fire exposure as established in accordance with the test procedures of NFPA 251, Standard Methods of Tests of Fire Endurance of Building Construction and Materials. NFPA 220, 1999 ed.

Fire Resistant. Construction designed to provide reasonable protection against fire. NFPA 495, 2001 ed.; NFPA 1141, 2003 ed.

Fire Retardant. A liquid, solid, or gas that tends to inhibit combustion when applied on, mixed in, or combined with combustible materials. NFPA 1, 2003 ed.

Fire Scene Reconstruction. The process of recreating the physical scene during fire scene analysis through the removal of debris and the replacement of contents or structural elements in their pre-fire positions. NFPA 921, 2004 ed.

Fire Science. The body of knowledge concerning the study of fire and related subjects (such as combustion, flame, products of combustion, heat release, heat transfer, fire and explosion chemistry, fire and explosion dynamics, thermodynamics, kinetics, fluid mechanics, fire safety) and their interaction with people, structures, and the environment. NFPA 921, 2004 ed.

Fire Service. Career or volunteer service groups that are organized and trained for the prevention and control of loss of life and property from any fire or disaster. NFPA 1404, 2002 ed.

Fire Service Personnel. All employees, whether career or volunteer, of a fire department who are assigned or may be assigned to perform duties at emergency incidents. NFPA 901, 2001 ed.

Fire Service Vehicle. Any vehicle operated by a fire department. NFPA 1451, 2002 ed.

Fire Shelter. An item of protective equipment configured as an aluminized tent utilized for protection, by means of reflecting radiant heat, in a fire entrapment situation. (See Figure F.10.) NFPA 1500, 2002 ed.

Fire Situation. Factors pertaining to a fire that affect decisions relating to fire suppression including, but not limited to, fuel types and geometry, fire characteristics and behavior, life safety hazard, type of structure, exposure, and weather. NFPA 1145, 2000 ed.

Fire Spread. The movement of fire from one place to another. NFPA 921, 2004 ed.

Fire Station. A location for the fire fighting water supply outlet, hose, and equipment on board ship. NFPA 1405, 2001 ed.

Figure F.10 Fire Shelter.

Fire Suppression. All the work of confining and extinguishing wildland fires. NFPA 1051, 2002 ed.

Fire Suppression. The activities involved in controlling and extinguishing fires. NFPA 1500, 2002 ed.

Fire Under Control. That point in time when a fire is sufficiently surrounded and quenched so that in the judgment of the commanding officer it no longer threatens destruction of additional property, or in wildland fire, that point in time when a control line is around a fire, any spot fires therefrom, and any interior islands to be saved. NFPA 901, 2001 ed.

Fire Wall. *As applied to aircraft:* A bulkhead designed to stop the lateral spread of fire in a fuselage or engine nacelle. NFPA 402, 2002 ed. *As applied to buildings:* A wall separating buildings or subdividing a building to prevent the spread of the fire and having a fire resistance rating and structural stability. (See Figure F.11.) NFPA 221, 2000 ed. *As applied to incident reporting:* A fire division assembly with a fire resistance rating of 3 test hours or longer, built to permit complete burnout and collapse of the structure on one side without extension of fire through the fire wall or collapse of the fire wall. NFPA 901, 2001 ed.

Fire Warp. Wire rope or other fireproof materials of sufficient strength to tow the vessel in the event of fire. NFPA 1405, 2001 ed.

Fire Watch. The assignment of a person or persons to an area for the express purpose of notifying the fire department, the building occupants, or both of an emergency; preventing a fire from occurring; extinguishing small fires; or protecting the public from fire or life safety dangers. NFPA 1, 2003 ed.

FIGURE F.11 Fire Wall.

Fire-Rated Assembly. An assembly (for example, wall, floor, or roof) that has been tested using standard test methods and has received at least a 1-hour fire resistance rating. NFPA 901, 2001 ed.

Fire-Resistant Construction. Construction designed to offer reasonable protection against fire. NFPA 1144, 2002 ed.

Firesetting. Any unsanctioned incendiary use of fire, including both intentional and unintentional involvement, whether or not an actual fire and/or explosion occurs. NFPA 1035, 2005 ed.

First Responder at the Awareness Level. Those persons who, in the course of their normal duties, could be the first on the scene of an emergency involving hazardous materials and who are expected to recognize the presence of hazardous materials, protect themselves, call for trained personnel, and secure the area. NFPA 472, 2002 ed.

First Responder at the Operational Level. Those persons who respond to releases or potential releases of hazardous materials as part of the initial response to the incident for the purpose of protecting nearby persons, the environment, or property from the effects of the release and who are expected to respond in a defensive fashion to control the release from a safe distance and keep it from spreading. NFPA 472, 2002 ed.

First Responder (EMS). Functional provision of initial assessment (i.e., airway, breathing, and circulatory systems) and basic first-aid intervention, including CPR and automatic external defibrillator (AED) capability. NFPA 1710, 2004 ed.

Fit. The quality, state, or manner in which the length and closeness of clothing, when worn, relates to the human body. NFPA 1851, 2001 ed.

Fixed Base Operator (FBO). An enterprise based on an airport that provides storage, maintenance, or service for aircraft operators. NFPA 403, 2003 ed.

Fixed Electrical Equipment. Any electrical equipment that is not removable without the use of tools or is hard wired to the vehicle's electrical system. NFPA 1901, 2003 ed.

Fixed Line (Fixed Line System). A rope rescue system consisting of a nonmoving rope attached to an anchor system. NFPA 1670, 2004 ed.

Fixed Object. An object, device, or appliance, for example, a steam radiator, that is fastened or secured at a specific location. NFPA 901, 2001 ed.

Fixed Power Source. Any line voltage power source other than a portable generator. NFPA 1901, 2003 ed.

Fixed Tank. A tank that is internal to or attached directly to a helicopter. NFPA 1150, 2004 ed.

Fixed-Temperature Detector. A device that responds when its operating element becomes heated to a predetermined level. NFPA 72, 2002 ed.

Flame. A body or stream of gaseous material involved in the combustion process and emitting radiant energy at specific wavelength bands determined by the combustion chemistry of the fuel. *Note: In most cases, some portion of the emitted radiant energy is visible to the human eye.* NFPA 72, 2002 ed.

Flame Break. A solid material, without holes or other openings, used to retard the spread of flame. NFPA 1124, 2003 ed.

Flame Front. The leading edge of burning gases of a combustion reaction. NFPA 921, 2004 ed.

Flame Propagation Rate. The speed at which a flame progresses through a combustible fuel-air mixture. NFPA 86, 2003 ed.

Flame Resistance. The property of a material whereby combustion is prevented, terminated, or inhibited following the application of a flaming or nonflaming source of ignition, with or without subsequent removal of the ignition source. *Note: Flame resistance can be an inherent property of a material, or it can be imparted by specific treatment.* NFPA 1500, 2002 ed.

Flame Speed. The rate of flame propagation relative to the velocity of the unburned gas that is ahead of it. NFPA 68, 2002 ed.

Flame Spread. The propagation of flame over a surface. NFPA 101, 2003 ed.

Flameover. The condition where unburned fuel (pyrolysate) from the originating fire has accumulated in the ceiling layer to a sufficient concentration (i.e., at or above the lower flammable limit) that it ignites and burns; can occur without ignition and prior to the ignition of other fuels separate from the origin. NFPA 921, 2004 ed.

Flames. Products of combustion that are illuminated by the heat of combustion and accompany the burning of most materials in normal atmospheres. NFPA 901, 2001 ed.

Flammable. A combustible (solid, liquid, or gas) that is capable of easily being ignited and rapidly consumed by fire. NFPA 1126, 2001 ed.

Flammable Limit. The upper or lower concentration limit at a specified temperature and pressure of a flammable gas or a vapor of an ignitable liquid and air, expressed as a percentage of fuel by volume that can be ignited. (See Figure F.12.) NFPA 921, 2004 ed.

Flammable Liquid. A liquid that has a closed-cup flash point that is below 37.8°C (100°F) and a maximum vapor pressure of 2068 mm Hg (40 psia) at 37.8°C (100°F). NFPA 30, 2003 ed.

Flammable or Explosive Atmospheres. Atmospheres containing solids, liquids, vapors, or gases at concentrations that will burn or explode if ignited. NFPA 1991, 2005 ed.

Flammable Range. The range of concentrations between the lower and upper flammable limits. (See Figure F.13.) NFPA 68, 2002 ed.

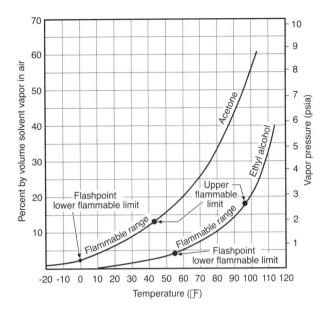

FIGURE F.13 Flammable Range.

Flare. A pyrotechnic device designed to produce a single source of intense light for a defined period of time. NFPA 1126, 2001 ed.

Flash Fire. A fire that spreads rapidly through a diffuse fuel, such as dust, gas, or the vapors of an ignitible liquid, without the production of damaging pressure. NFPA 921, 2004 ed.

Flash Point. The minimum temperature at which a liquid or a solid emits vapor sufficient to form an ignitable mixture with air near the surface of the liquid or the solid. (See Figure F.14.) NFPA 30, 2003 ed.

Flash Pot. A device used with flashpowder that produces a flash of light and is capable of directing the flash in an upward direction. NFPA 1126, 2001 ed.

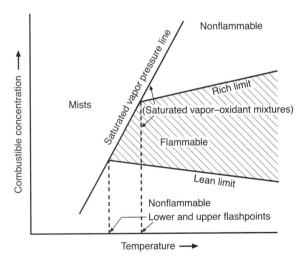

FIGURE F.12 Flammable Limit.

FIGURE F.14 Flash Point of Gasoline.

Flashback. The tendency of flammable liquid fires to re-ignite from any source of ignition after the fire has once been extinguished. NFPA 402, 2002 ed.

Flashover. A transition phase in the development of a compartment fire in which surfaces exposed to thermal radiation reach ignition temperature more or less simultaneously and fire spreads rapidly throughout the space, resulting in full room involvement or total involvement of the compartment or enclosed space. NFPA 921, 2004 ed.

Fleet Vehicle Motor Fuel Dispensing Facility. A motor fuel dispensing facility at a commercial, industrial, governmental, or manufacturing property where motor fuels are dispensed into the fuel tanks of motor vehicles that are used in connection with the business or operation of that property by persons within the employ of such business or operation. NFPA 30A, 2003 ed.

Flight Attendant. A member of the flight deck crew whose responsibility includes the management of activities within the passenger cabin. NFPA 402, 2002 ed.

Flight Data Recorder (FDR). An instrument that monitors performance characteristics of an aircraft in flight. *Note: A flight data recorder is usually mounted in the tail area of an aircraft and is designed to withstand certain impact forces and a degree of fire. Its purpose is to provide investigators with flight performance data that might be relevant in determining the cause of an accident/incident.* NFPA 402, 2002 ed.

Flight Deck Crew. Those members of the crew whose responsibility includes the management of the aircraft's flight control and ground movements. NFPA 402, 2002 ed.

Flight Service Station (FSS). An air traffic facility that briefs pilots, processes, and monitors flight plans, and provides in-flight advisories. NFPA 403, 2003 ed.

Flight Technical Crew (FTC). Includes pilots, flight engineers, and flight attendants who crew on aircraft movement. NFPA 402, 2002 ed.

Flood Insurance Rate Maps. Maps produced by the National Flood Insurance Program, under the auspices of the Federal Emergency Management Agency (FEMA), that illustrate geographic areas that are subject to flooding. NFPA 1006, 2003 ed.

Flotation Aids. Devices that provide supplemental flotation for persons in the water but do not meet U.S. Coast Guard performance criteria such as breaking strength of the thread used in sewing the device, the usable life of the flotation materials including compressibility factors, the colors and fading potential of certain dyes used in the fabrication of the device, and the strength and breaking force required for buckles and tie straps. NFPA 1006, 2003 ed.

Flow Chart. A diagram consisting of a set of symbols and connecting lines that shows a step-by-step progression through a procedure or system. NFPA 1401, 2001 ed.

Fluid Coupling. A turbine-like device that transmits power solely through the action of a fluid in a closed circuit without direct mechanical connection between input and output shafts and without producing torque multiplication. NFPA 414, 2001 ed.

Fluid-Resistant Clothing. Clothing worn for the purpose of isolating parts of the wearer's body from contact with body fluids. NFPA 1581, 2005 ed.

Fluorescence. A process by which radiant flux of certain wavelengths is absorbed and reradiated nonthermally in other, usually longer, wavelengths. NFPA 1976, 2000 ed.

Fluoroprotein Foam (FP). A protein-based foam concentrate with added fluorochemical surfactants that forms a foam showing a measurable degree of compatibility with dry chemical extinguishing agents and an increase in tolerance to contamination by fuel. NFPA 403, 2003 ed.

Flush. A nozzle feature that allows the orifice to be opened so that small debris that could otherwise be trapped in the nozzle, causing pattern disruptions and discharge variation, can pass through. NFPA 1964, 2003 ed.

Fly Section. Any section of an aerial telescoping device beyond the base section. NFPA 1901, 2003 ed.

Fly Section(s). The upper section(s) of an extension ladder. NFPA 1931, 2004 ed.

FMVSS. See Federal Motor Vehicle Safety Standard.

Foam Application Rate. The amount of foam solution in liters or gallons per minute expressed as a relationship with a unit of area, usually square meter or square foot. NFPA 402, 2002 ed.

Foam Blanket. A covering of foam over the surface of flammable liquids to provide extinguishment and prevent ignition. NFPA 402, 2002 ed.

Foam Burnback Resistance. The ability of a foam blanket to retain aerated moisture and resist destruction by heat and flame. NFPA 402, 2002 ed.

Foam Chemical. A generic name for the liquid foam concentrate, foam solution, and foam in whatever form it is being used. NFPA 1150, 2004 ed.

Foam Concentrate. A concentrated liquid foaming agent as received from the manufacturer. NFPA 16, 2003 ed.

Foam Concentrate Proportioning System. The foaming agent as received from the supplier that, when mixed with water, becomes foam solution. NFPA 1145, 2000 ed.

Foam Drain Time. The time required for 25 percent of the original foam solution (foam concentrate plus water) to drain out of the foam. *Note: Foam drain time is also known as the "25 percent drainage time" or "drainage time."* NFPA 402, 2002 ed.

Foam Expansion. The ratio between the volume of foam produced and the volume of solution used in its production. NFPA 412, 2003 ed.

Foam Pattern. The ground area over which foam is distributed during the discharge of a foam-making device. NFPA 412, 2003 ed.

Foam Proportioner. A device or method to add foam concentrate to water to make foam solution. NFPA 1901, 2003 ed.

Foam Proportioning System. The apparatus and techniques used to mix concentrate with water to make foam solution. NFPA 1901, 2003 ed.

Foam Stability. The degree to which a foam resists spontaneous collapse or degradation caused by external influences such as heat or chemical action. NFPA 412, 2003 ed.

Foam System. A system provided on fire apparatus for the delivery of a proportioned foam and water mixture for use in fire extinguishment. *Note: A foam system includes a concentrate tank, a method for removing the concentrate from the tank, a foam-liquid proportioning system, and a method (e.g., hand lines or fixed turret nozzles) of delivering the proportioned foam to the fire.* NFPA 1002, 2003 ed.

Foam Weep. That portion of foam that is separated from the principal foam stream during discharge and falls at short range. NFPA 412, 2003 ed.

Foam-Liquid Concentrate Percentage. The percentage of foam-liquid concentrate in solution with water. NFPA 414, 2001 ed.

Foam-Producing System. A collection of equipment that, by various means, combines water and a controlled amount of foam concentrate into a foam solution and then aerates that solution. NFPA 1145, 2000 ed.

Fold. A transverse bend (fold) occurring where the hose is lengthwise doubled over on itself, as on a pin rack. NFPA 1962, 2003 ed.

Follow-Up Program. The sampling, inspections, test, or other measures conducted by a certification organization on a periodic basis to determine the continued compliance of labeled and listed products that are being produced by the manufacturer to the requirements of a standard. NFPA 1977, 2005 ed.

Foodborne Pathogen. A microorganism that is present in food or drinking water and can cause infection and/or disease in humans. NFPA 1581, 2005 ed.

Footwear Upper. That portion of the footwear element above the sole, heel, and insole. NFPA 1991, 2005 ed.

Force Multiplier. Any load, object, environmental factor, or system configuration that increases the load on the anchor system(s). NFPA 1670, 2004 ed.

Forcible Entry. Techniques used by fire personnel to gain entry into buildings, vehicles, aircraft, or other areas of confinement when normal means of entry are locked or blocked. (See Figure F.15.) NFPA 1710, 2004 ed.

Forecastle (fo'c's'le). (1) The section of the upper deck of a ship located at the bow, forward of the foremast. (2) A superstructure at the bow of a ship where maintenance shops, rope lockers, and paint lockers are located. NFPA 1405, 2001 ed.

Figure F.15 Forcible Entry.

Forensic. Legal; pertaining to courts of law. NFPA 921, 2004 ed.

Forestry Fire Hose. A hose designed to meet specialized requirements for fighting wildland fires. NFPA 1961, 2002 ed.

Forward (Fore). The direction toward the bow of the vessel. NFPA 1405, 2001 ed.

Forward-Looking Infrared (FLIR). The detection of heat energy radiated by objects to produce a "thermal image," which is converted by electronics and signal processing into a visual image that can be viewed by the operator. (See Figure F.16.) NFPA 414, 2001 ed.

Fractile Response Interval. A method of describing response intervals that uses frequency distribution as its basis for reporting. NFPA 450, 2004 ed.

Fracture. A type of defect found in welds that has a large length-to-width ratio and travels through or adjacent to the metal grain boundaries. *Note: Usually, this type of defect is referred to as a crack.* NFPA 1914, 2002 ed.

Frame. The basic structural system that transfers the weight of the fire apparatus to the suspension system. NFPA 1915, 2000 ed.

Frangible Gate/Fence. Gates or fence sections designed to open, break away, or collapse when struck with the bumper of an ARFF vehicle responding to an emergency. NFPA 402, 2002 ed.

Free Weight. Test weights that are not controlled from any direction except by the force of gravity. NFPA 1932, 2004 ed.

Freeboard. The vertical distance between the sheer and the waterline measured at a stated point on the length of the vessel or at the lowest point of the sheer, and at designated displacement in fresh water. NFPA 1925, 2004 ed.

Figure F.16 Thermal Imaging Device.

Frequency. The number of occurrences per unit time at which observed events occur or are predicted to occur. NFPA 1250, 2004 ed.

Fresh Water. Deionized or distilled water to which 140 ppm of calcium chloride has been added. NFPA 1150, 2004 ed.

Front Length. Upper torso garment measurement from bottom collar seam to the bottom edge of the garment at front edge. NFPA 1977, 2005 ed.

Front Rise. Lower torso garment measurement from crotch seam to top of waistband at front center. NFPA 1977, 2005 ed.

Front Waist Pocket(s). Slanted or side seam opening pockets that open to the exterior, located at or near the front waist of a garment. NFPA 1977, 2005 ed.

Fuel. *As applied to oxygen-enriched atmospheres:* Any material that will maintain combustion under specified environmental conditions. NFPA 53, 2004 ed. *As applied to pyrotechnics:* Anything combustible or acting as a chemical-reducing agent such as, but not limited to, sulfur; aluminum powder; iron powder; charcoal; magnesium; gums; and organic plastic binders. *Note: Fuels are an ingredient of pyrotechnic materials.* NFPA 1126, 2001 ed. *As applied to wildland fire fighting:* All combustible material within the wildland/urban interface or intermix including, but not limited to, vegetation and structures. NFPA 1144, 2002 ed.

Fuel Load. The total quantity of combustible contents of a building, space, or fire area, including interior finish and trim, expressed in heat units or the equivalent weight in wood. NFPA 921, 2004 ed.

Fuel Modification. Any manipulation or removal of fuels to reduce the likelihood of ignition or the resistance to fire control. NFPA 1144, 2002 ed.

Fuel Servicing. Fueling and defueling of aircraft fuel tanks, not including aircraft fuel transfer operations and design of aircraft fuel systems during aircraft maintenance or manufacturing operations. NFPA 402, 2002 ed.

Fuel-Controlled Fire. A fire in which the heat release rate and growth rate are controlled by the characteristics of the fuel, such as quantity and geometry, and in which adequate air for combustion is available. NFPA 921, 2004 ed.

Full Face Mask. A diving mask that covers the diver's entire face, includes a regulator for breathing, has separate inhalation and exhalation chambers, provides for defogging, permits free flow if the seal is broken, and provides for a communication module. NFPA 1670, 2004 ed.

Full Room Involvement. Condition in a compartment fire in which the entire volume is involved in fire. NFPA 921, 2004 ed.

Full-Time Swivel. A connection that allows one side of the connection to swivel or rotate in relation to the other side after the connection has been tightened together. NFPA 1964, 2003 ed.

Fully Charged. An SCBA cylinder filled to the SCBA manufacturer's specified pressure level. NFPA 1852, 2002 ed.

Fully Enclosed Area. A cab or passenger compartment of fire apparatus providing total enclosure equipped with positive latching doors for entry and exit. NFPA 1500, 2002 ed.

Fully Enclosed Personnel Area. A driver or passenger compartment on the fire apparatus that provides total enclosure on all sides, top, and bottom and has positive latching on all access doors. NFPA 1901, 2003 ed.

Fully Loaded Vehicle. Consists of the fully assembled vehicle, complete with a full complement of crew, fuel, and fire fighting agents. NFPA 414, 2001 ed.

Functional/Functionality. The ability of an element or component of an element to continue to be used for its intended purpose. NFPA 1971, 2000 ed.

Functional Capability. A learned ability involving skills of specialized activities. NFPA 1951, 2005 ed.

Functional Capacity Evaluation. An assessment of the correlation between that individual's capabilities and the essential job tasks. NFPA 1582, 2003 ed.

Function. One of a group of related actions contributing to the overall goals of the organization. NFPA 1401, 2001 ed.

Fuselage. The main body of an aircraft. NFPA 402, 2002 ed.

Gallon. U.S. Standard. 1 U.S. gal = 0.833 Imperial gal = 231 in. 3 = 3.785 L. NFPA 58, 2004 ed.

Galvanic Corrosion. The corrosion that occurs at the anode of a galvanic couple caused by the flow of ions between dissolution metals in an electrolyte and electron flow within the dissimilar metals. NFPA 1925, 2004 ed.

Gangway. The opening through bulwarks (sides) of a ship or a ship's rail to which an accommodation ladder used for normal boarding of the ship is attached. NFPA 1405, 2001 ed.

Garage. A building or portion of a building in which one or more self-propelled vehicles carrying volatile flammable liquid for fuel or power are kept for use, sale, storage, rental, repair, exhibition, or demonstrating purposes, and all that portion of a building that is on or below the floor or floors in which such vehicles are kept and that is not separated there by suitable cutoffs. NFPA 5000, 2002 ed.

Garment Closure. The garment component designed and configured to allow the wearer to enter (don) and exit (doff) the garment. NFPA 1999, 2003 ed.

Garment Closure Assembly. The combination of the garment closure and the seam attaching the garment closure to the garment, excluding any protective flap or cover. NFPA 1999, 2003 ed.

Garment Material. The principal chemical-protective material used in the construction of the liquid splash-protective suit. NFPA 1992, 2005 ed.

Garment(s). The coat, trouser, or coverall elements of the protective ensemble designed to provide minimum protection to the upper and lower torso, arms, and legs, excluding the head, hands, and feet. NFPA 1851, 2001 ed.

Gas. The state of matter characterized by complete molecular mobility and unlimited expansion; used synonymously with the term vapor. NFPA 68, 2002 ed.

Gas Cabinet. A fully enclosed, noncombustible enclosure used to provide an isolated environment for compressed gas cylinders in storage or use. NFPA 5000, 2002 ed.

Gas Manufacturer/Producer. A business that produces compressed gases or cryogenic fluids, or both, or fills portable or stationary gas containers, cylinders, or tanks. NFPA 55, 2005 ed.

Gauge. A round, analog pressure-indicating device that uses mechanical means to measure pressure. (*See Figure G.1.*) NFPA 1911, 2002 ed.

Gauge Pressure. Pressure measured by an instrument where the pressure indicated is relative to atmospheric pressure. NFPA 1911, 2002 ed.

Gauntlet. A glove term for the circular, flared, or otherwise expanded part of the glove that extends beyond the opening of the glove body. NFPA 1971, 2000 ed.

GAWR. See Gross Axle Weight Rating.

GCWR. See Gross Combination Weight Rating.

General Area. An area surrounding the incident site (e.g., collapsed structure or trench) that has a size proportional to the size and nature of the incident. *Note: The general area is also referred to as the "warm zone." Within the general area, access by people, heavy machinery, and vehicles is limited and strictly controlled.* NFPA 1006, 2003 ed.

Figure G.1 Gauge.

General Property Use. The actual general (overall) use of land or space under the same management or ownership, or within the same legal boundaries, including any structures, vehicles, or other appurtenances thereon. NFPA 901, 2001 ed.

General Staff. Responders that serve as section chiefs of operations, planning, logistics, and finance/administration. NFPA 1561, 2005 ed.

General-Purpose Warehouse. A separate, detached building or portion of a building used only for warehousing-type operations. NFPA 30, 2003 ed.

Generator. A device that develops either direct or alternating electrical voltage at any frequency. (See Figure G.2.) NFPA 1221, 2002 ed.

Generator (Alternator), Fixed. A mechanically driven electrical source, usually 7 kW or greater, that is permanently secured to the vehicle. NFPA 1901, 1999 ed.

Geographic Information System (GIS). A system of computer software, hardware, data, and personnel to describe information tied to a spatial location. NFPA 450, 2004 ed.

Gerb. A cylindrical preload intended to produce a controlled spray of sparks with a reproducible and predictable duration, height, and diameter. NFPA 1126, 2001 ed.

Global Positioning System (GPS). A satellite-based radio navigation system comprised of three segments: space, control, and user. NFPA 414, 2001 ed.

Glove Body. The part of the glove that extends from the tip of the fingers to 25 mm (1 in.) beyond the wrist crease. NFPA 1971, 2000 ed.

Glove Gauntlet. The circular, flared, or otherwise expanded part of the glove that extends beyond the opening of the glove body. NFPA 1951, 2005 ed.

Glove Liner. The innermost component of the glove body composite that comes into contact with the wearer's skin. NFPA 1971, 2000 ed.

Glove Material. All material layers used in the construction of gloves. NFPA 1999, 2003 ed.

Glove Wristlet. The circular, close-fitting part of the glove, usually made of knitted material, that extends beyond the opening of the glove body. NFPA 1851, 2001 ed.

Gloves. An element of the protective ensemble designed to provide minimum protection to the fingers, thumb, hand, and wrist. NFPA 1971, 2000 ed.

Glowing Combustion. Luminous burning of solid material without a visible flame. NFPA 921, 2004 ed.

GM. Abbreviation of metacentric height. NFPA 1925, 2004 ed.

Goggle. The helmet component, not intended as primary eye protection, but intended to help protect the wearer's eyes and a portion of the wearer's face. NFPA 1976, 2000 ed.

Goggle Clip. The component of the helmet that retains the strap of the goggles or headlamp. NFPA 1977, 2005 ed.

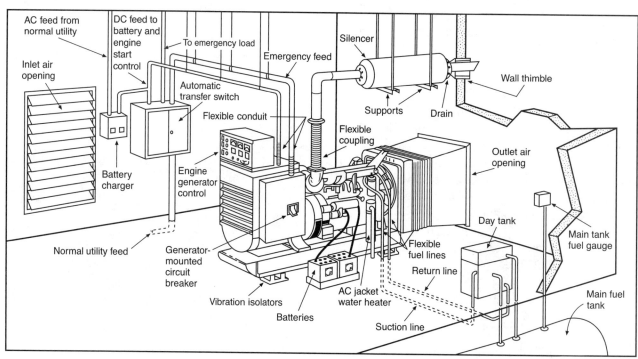

FIGURE G.2 Generator.

gpm. Gallons per minute. NFPA 1901, 1999 ed.

Grade. A measurement of the angle used in road design and expressed as a percentage of elevation change over distance. NFPA 1901, 2003 ed.

Grade Pole. A wood or fiberglass pole, either cut to a certain length or provided with markings, used by workers when setting pipes on grade. NFPA 1670, 2004 ed.

Grading. The process of proportioning components for construction of an element. NFPA 1976, 2000 ed.

Grid Map. A plan view of an area with a system of squares (numbered and lettered) superimposed to provide a fixed reference to any point in the area. NFPA 402, 2002 ed.

Gross Axle Weight Rating (GAWR). The chassis manufacturer's specified maximum load-carrying capacity of an axle system, as measured at the tire–ground interfaces. NFPA 1901, 2003 ed.

Gross Combination Weight Rating (GCWR). The chassis manufacturer's specified maximum load-carrying capacity for tractor trailer-type vehicles having three or more axle systems. *Note: A multiaxle installation is one system.* NFPA 1901, 2003 ed.

Gross Decontamination. The initial phase of the decontamination process during which the amount of surface contaminant is significantly reduced. (See Figure G.3.) NFPA 471, 2002 ed.

Gross Floor Area. The floor area within the inside pcrimeter of the outside walls of the building under consideration with no deduction for hallways, stairs, closets, thickness of interior walls, columns, or other features. NFPA 5000, 2002 ed.

Gross Vehicle Weight Rating (GVWR). The chassis manufacturer's specified maximum load-carrying capacity of a vehicle having two axle systems. *Note: A multiaxle installation is one system.* NFPA 1901, 2003 ed.

Ground Clearance. The clearance under a vehicle at all locations except the axles and driveshaft connections to the axle. NFPA 1906, 2001 ed.

Ground Elevation, Adjacent. The reference plane representing the average elevation of the finished ground level measured at a distance of 3 m (10 ft) from all exterior walls of the building. NFPA 1141, 2003 ed.

Ground Fault. An unintended current that flows outside the normal circuit path, such as (a) through the equipment grounding conductor, (b) through conductive material in contact with lower potential (such as earth), other than the electrical system ground (metal water or plumbing pipes, etc.), (c) through a combination of these ground return paths. NFPA 921, 2004 ed.

Figure G.3 Gross Decontamination.

Ground Fuels. All combustible materials such as grass, duff, loose surface litter, tree or shrub roots, rotting wood, leaves, peat, or sawdust that typically support combustion. NFPA 1144, 2002 ed.

Ground Kettle. A container that could be mounted on wheels and is used for heating tar, asphalt, or similar substances. NFPA 1, 2003 ed.

Ground Sweep Nozzle. A small nozzle(s) mounted in front of the vehicle that disperses foam solution in front to provide protection. NFPA 414, 2001 ed.

Ground Tackle. A general term for the anchor, anchor rodes, and fittings used for securing a vessel to anchor. NFPA 1925, 2004 ed.

Grounded Conductor. In marine fire fighting vessels, a current-carrying conductor connected to the side of the electrical source that is intentionally maintained at ground potential. NFPA 1925, 2004 ed.

Grounding Conductor. In marine fire fighting vessels, a normally non-current-carrying conductor provided to connect the exposed metallic enclosures of electrical equipment to ground for the purpose of minimizing shock hazard to personnel. NFPA 1925, 2004 ed.

Group. A supervisory level established to divide the incident into functional areas of operation. NFPA 1561, 2005 ed.

GSA KKK Specification. A set of federal specifications relating to purchasing requirements for ambulance design and manufacture. NFPA 450, 2004 ed.

Gunwale. The upper edge of a side of a vessel or boat designed to prevent items from being washed overboard. NFPA 1405, 2001 ed.

GVWR. See Gross Vehicle Weight Rating.

H

Halogenated Agent. A liquefied gas extinguishing agent that extinguishes fire by chemically interrupting the combustion reaction between fuel and oxygen. *Note: Halogenated agents leave no residue.* NFPA 402, 2002 ed.

Halon 1211. A halogenated agent whose chemical name is bromochlorodifluoromethane, $CBrClF_2$, and that is a multipurpose, Class ABC-rated agent effective against flammable liquid fires. NFPA 402, 2002 ed.

Halon 1301. A halogenated agent whose chemical name is bromotrifluoromethane, $CBrF_3$, and that is recognized as an agent having Class ABC capability in total flooding systems. NFPA 402, 2002 ed.

Halyard. Rope used on extension ladders for the purpose of raising a fly section(s). (See Figure H.1.) NFPA 1931, 2004 ed.

Handhold Device or Grab Rail. Any fitting, assembly, or device, other than a lifeline or deck rail, that is intended for grasping with the hand. *Note: The handhold device can be of metal, wood, plastic, reinforced fiberglass, or any combination of materials suited for the purpose.* NFPA 1925, 2004 ed.

Handline Nozzle. A nozzle with a rated discharge of less than 1325 L/min (350 gpm). NFPA 1964, 2003 ed.

Handwashing Facility. A facility providing an adequate supply of running potable water, soap, and single-use towels or hot-air drying machines. NFPA 1581, 2005 ed.

Hard Suction Hose. A hose used for drafting water from static supplies (lakes, rivers, wells, etc.). (See Figure H.2.) It is also used for supplying pumpers from a hydrant if designed for that purpose. *Note: The hose contains a semirigid or rigid reinforcement designed to prevent collapse of the hose under vacuum.* NFPA 1963, 1998 ed.

Hardware. *As applied to structural fire fighting:* Nonfabric components of the proximity protective ensemble including, but not limited to, those made of metal or plastic. NFPA 1971, 2000 ed. *As applied to search and rescue operations:* Rigid mechanical auxiliary equipment that can include, but is not limited to, anchor plates, carabiners, and mechanical ascent and descent control devices. NFPA 1670, 2004 ed.

Hauling System. A rope system generally constructed from life safety rope, pulleys, and other rope rescue system components capable of lifting or moving a load across a given area. NFPA 1006, 2003 ed.

Hawse Pipe. A cylindrical or elliptical pipe or casting in a vessel's hull through which the anchor rode runs and within which the anchor shank can be housed. NFPA 1925, 2004 ed.

FIGURE H.1 Halyard.

FIGURE H.2 Hard Suction Hose.

Hazard. *As applied to fire and explosion investigation:* Any arrangement of materials and heat sources that presents the potential for harm, such as personal injury or ignition of combustibles. NFPA 921, 2004 ed. *As applied to industrial machinery:* A source of possible injury or damage to health. NFPA 79, 2002 ed. *As applied to Emergency Service Organization:* A condition, situation, attitude, or action that creates or increases expected loss frequency or severity. NFPA 1250, 2004 ed.

Hazard Assessment System. A system to evaluate and rate pertinent factors such as fire and weather history, fuels, improvements, topography, and access to develop and implement mitigation strategies. NFPA 1051, 2002 ed.

Hazard Identification. The process of identifying situations or conditions that have the potential to cause injury to people, damage to property, or damage to the environment. NFPA 1670, 2004 ed.

Hazard Mitigation. Activities taken to isolate, eliminate, or reduce the degree of risk to life and property from hazards, either before, during, or after an incident. NFPA 1006, 2003 ed.

Ordinary Hazard. Those hazards that are likely to burn with moderate rapidity or to give off a considerable volume of smoke. NFPA 1, 2000 ed.

Hazard Rating. The numerical rating of the health, flammability, and self-reactivity, and other hazards of the material, including its reaction with water, as specified in NFPA 704, Standard System for the Identification of the Hazards of Materials for Emergency Response. (See Figure H.3.) NFPA 55, 2005 ed.

Hazard Sector. That function within an overall incident management system that deals with the mitigation of a hazardous materials incident. NFPA 471, 2002 ed.

Hazard/Hazardous. Capable of posing an unreasonable risk to health, safety, or the environment; capable of causing harm. NFPA 471, 2002 ed.

Hazardous Atmosphere. *As applied to fire service training:* Any atmosphere that is oxygen deficient or that contains a toxic or disease-producing contaminant. NFPA 1404, 2002 ed. *As applied to search and rescue operations:* Any atmosphere that can expose personnel to the risk of death, incapacitation, injury, acute illness, or impairment of the ability to self-rescue. NFPA 1670, 2004 ed.

Hazardous Material. A substance (solid, liquid, or gas) that when released is capable of creating harm to people, the environment, and property. (See Figure H.4.) NFPA 472, 2002 ed.

Hazardous Material Response Fire Apparatus. An emergency vehicle designed to carry various support equipment and personnel to a scene of a hazardous material incident. NFPA 1901, 2003 ed.

Hazardous Material Storage Facility. A building, a portion of a building, or exterior area used for the storage of hazardous materials in excess of exempt amounts. NFPA 1, 2003 ed.

Hazardous Materials Branch. That function within an overall incident management system that deals with the mitigation of the hazardous materials portion of a hazardous materials incident. NFPA 472, 2002 ed.

FIGURE H.3 Hazard Rating.

FIGURE H.4 Hazardous Material.

Hazardous Materials Branch Officer. The person who is responsible for directing and coordinating all operations assigned to the hazardous materials branch by the incident commander. NFPA 472, 2002 ed.

Hazardous Materials Branch Safety Officer. The person who works within an incident management system (IMS) to ensure that recognized safe practices are followed within the hazardous materials branch. NFPA 472, 2002 ed.

Hazardous Materials Emergency. An incident involving the release or potential release of hazardous chemicals into the environment that can cause loss of life, personal injury, or damage to property and the environment. NFPA 1971, 2000 ed.

Hazardous Materials Operations. All activities performed at the scene of a hazardous materials incident that expose fire department members to the dangers of hazardous materials. NFPA 1500, 2002 ed.

Hazardous Materials Response Team. An organized group of trained response personnel operating under an emergency response plan and appropriate standard operating procedures who handle and control actual or potential leaks or spills of hazardous materials requiring possible close approach to the material. NFPA 472, 2002 ed.

Hazardous Materials Sector Officer. The person responsible for the management of the hazard sector. NFPA 471, 2002 ed.

Hazardous Materials Storage Locker. A movable prefabricated structure, manufactured primarily at a site other than the final location of the structure and transported completely assembled or in a ready-to-assemble package to the final location. *Note: It is intended to meet local, state, and federal requirements for outside storage of hazardous materials.* NFPA 30, 2003 ed.

Hazardous Materials Technician. Person who responds to releases or potential releases of hazardous materials for the purpose of controlling the release using specialized protective clothing and control equipment. NFPA 472, 2002 ed.

Hazardous Materials Technician with a Cargo Tank Specialty. Person who provides support to the hazardous materials technician, provides oversight for product removal and movement of damaged cargo tanks, and acts as a liaison between technicians and other outside resources. NFPA 472, 2002 ed.

Hazardous Materials Technician with a Tank Car Specialty. Person who provides support to the hazardous materials technician, provides oversight for product removal and movement of damaged tank cars, and acts as a liaison between technicians and other outside resources. NFPA 472, 2002 ed.

Hazardous Materials Technician with an Intermodal Tank Specialty. Person who provides support to the hazardous materials technician, provides oversight for product removal and movement of damaged intermodal tanks, and acts as a liaison between technicians and other outside resources. NFPA 472, 2002 ed.

Hazardous Reaction or Hazardous Chemical Reaction. Reactions that result in dangers beyond the fire problems relating to flash point and boiling point of either the reactants or of the products. NFPA 30, 2003 ed.

Haze. Light that is scattered as a result of passing through a transparent object. NFPA 1981, 2002 ed.

HBV. Hepatitis B virus. NFPA 1581, 2005 ed.

HCV. Hepatitis C virus. NFPA 1581, 2005 ed.

Headband. The portion of the helmet suspension that encircles the head. NFPA 1971, 2000 ed.

Headform. A device that simulates the configuration of the human head. NFPA 1971, 2000 ed.

Heads Up Display (HUD). Visual display of information and system condition status visible to the SCBA wearer. NFPA 1981, 2002 ed.

Health and Fitness Coordinator. A person who, under the supervision of the fire department physician, has been designated by the department to coordinate and be responsible for the health and fitness programs of the department. NFPA 1500, 2002 ed.

Health and Safety Committee. A representative group of individuals who serve along with the fire department physician and health and fitness coordinator. It is chaired by the fire department health and safety officer, who oversees the implementation of the fire department occupational safety and health program. NFPA 1582, 2003 ed.

Health and Safety Officer. The member of the fire department assigned and authorized by the fire chief as the manager of the health and safety program. NFPA 1500, 2002 ed.

Health Care Occupancy. An occupancy used for purposes of medical or other treatment or care of four or more persons where such occupants are mostly incapable of self-preservation due to age, physical or mental disability, or because of security measures not under the occupants' control. NFPA 5000, 2002 ed.

Health Data Base. A compilation of records and data that relates to the health experience of a group of individuals and is maintained in a manner such that it is retrievable for study and analysis over a period of time. NFPA 1500, 2002 ed.

Health Hazard. Any property of a material that either directly or indirectly can cause injury or incapacitation, either temporary or permanent, from exposure by contact, inhalation, or ingestion. NFPA 1521, 2002 ed.

Health Hazard Material. A chemical or substance classified as a toxic, highly toxic, or corrosive material. NFPA 5000, 2002 ed.

Health Maintenance Organization (HMO). An organized system of health care that provides or arranges for a range of basic and supplemental health care services to a voluntarily enrolled group of persons under a prepayment plan. NFPA 450, 2004 ed.

Health-Related Fitness Programs (HRFP). A comprehensive program designed to promote the member's ability to perform occupational activities with vigor and to assist the member in the attainment and maintenance of the premature development of injury, morbidity, and mortality. NFPA 1582, 2003 ed.

Heat. A form of energy characterized by vibration of molecules and capable of initiating and supporting chemical changes and changes of state. NFPA 921, 2004 ed.

Heat and Flame Vector. An arrow used in a fire scene drawing to show the direction of heat, smoke, or flame flow. NFPA 921, 2004 ed.

Heat Detector. A fire detector that detects either abnormally high temperature or rate of temperature rise, or both. (See Figure H.5.) NFPA 72, 1999 ed.

Heat Flux. The measure of the rate of heat transfer to a surface, expressed in kilowatts/m^2, kilojoules/m^2·s, or Btu/ft^2·s. NFPA 921, 2004 ed.

Heat of Ignition. The heat energy that brings about ignition. *Note: Heat energy comes in various forms and usually from a specific object or source. Therefore, the heat of ignition is divided into two parts: (a) equipment involved in ignition and (b) form of heat of ignition.* NFPA 901, 2001 ed.

Heat Release Rate (HRR). The rate at which heat energy is generated by burning. NFPA 921, 2004 ed.

Heat Resistance. The property of a foam to withstand exposure to high heat fluxes without loss of stability. NFPA 412, 2003 ed.

Heat Sensor Label. A label that changes color at a preset temperature to indicate a specific heat exposure. NFPA 1931, 2004 ed.

Heavy Construction Type. Construction that utilizes masonry, steel, and concrete in various combinations, including tilt-up, steel frame with infill, concrete moment resisting frame, concrete shearwall, unreinforced masonry infill in concrete frame, and precast concrete. NFPA 1006, 2003 ed.

Heavy Equipment. *As applied to wildland fire fighting:* Ground vehicles used in the suppression of wildland fires, such as dozers, tractors, plows, and their transport vehicles. *Note: Heavy equipment does not include fire apparatus.* NFPA 1051, 2002 ed. *As applied to rescue operations:* Typically, construction equipment that can include but is not limited to backhoes, trac hoes, grade-alls, and cranes. NFPA 1006, 2003 ed.

Heavy Load. Any load over 3175.15 kg (7000 lb). NFPA 1006, 2003 ed.

Heavy Object. An item of such size and weight that it cannot be moved without the use of power tools (e.g., hydraulic lifting devices) or complex mechanical advantage systems. NFPA 1670, 2004 ed.

Heavy Structural Collapse. Collapse of heavy construction–type buildings that require special tools and training to gain access into the building. NFPA 1006, 2003 ed.

Heel Breast. The forward face of the footwear heel. NFPA 1977, 2005 ed.

Heeling. (1) Tipping to one side. (2) Causing a ship to list. NFPA 1405, 2001 ed.

Height. As applied to a building, the vertical distance from the adjacent ground elevation to the average elevation of the roof of the highest story. NFPA 1141, 2003 ed.

Helm. The position that controls the direction and water speed of the vessel. *Note: The primary helm can be independent or located on the bridge. Secondary helms can be located for improved visibility for operations such as docking and towing.* NFPA 1925, 2004 ed.

Helmet. An element of the protective ensemble designed to provide minimum protection to the head. (See Figure H.6.) NFPA 1851, 2001 ed.

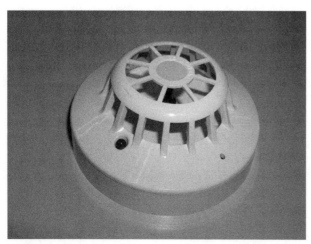

Figure H.5 Heat Detector.

FIGURE H.6 Helmet.

Helmet Outer Cover. A removable helmet component that offers radiant reflective protection to the exterior of the helmet shell. NFPA 1971, 2000 ed.

Helmet Positioning Index. The vertical distance, as specified by the helmet manufacturer, from the lowest point of the brow at the lateral midpoint of the helmet to the basic plane of the ISO Size J headform when the helmet is firmly positioned on the headform. NFPA 1971, 2000 ed.

Helmet Shell. A helmet without the suspension system, accessories, and fittings. NFPA 1977, 2005 ed.

Helmet Shroud. A component of the helmet element of the proximity protective ensemble designed to provide radiant reflective heat protection for the head and neck area. NFPA 1971, 2000 ed.

High Angle. Refers to an environment in which the load is predominantly supported by the rope rescue system. NFPA 1670, 2004 ed.

High Explosive. A material that is capable of sustaining a reaction front that moves through the unreacted material at a speed equal to or greater than that of sound in that medium [typically 1000 m/s (3000 ft/sec)]; a material capable of sustaining a detonation. NFPA 921, 2004 ed.

High Hazard. Those hazards likely to burn with extreme rapidity or from which explosions are likely. NFPA 520, 2005 ed.

High Hazard Level 1 Contents. Materials that present a detonation hazard including, but not limited to: explosives; organic peroxides, unclassified detonable; oxidizers, Class 4; pyrophoric materials, detonable; and unstable (reactive) materials, Class 3 detonable, and Class 4. NFPA 1, 2003 ed.

High Hazard Level 2 Contents. Materials that present a deflagration hazard or a hazard from accelerated burning including, but not limited to: Class I, II, or III-A flammable or combustible liquids that are used or stored in normally open containers or systems, or in closed containers or systems pressurized at more than 15 pounds per square inch (103.3kPa) gauge; combustible dusts stored, used, or generated in a manner creating a severe fire or explosion hazard; flammable gases and flammable cryogenic liquids; organic peroxides, Class I; oxidizers, Class 3 that are used or stored in normally open containers or systems, or in closed containers or systems pressurized at more than 15 pounds per square inch (103.2kPa) gauge; pyrophoric materials, nondetonable; unstable (reactive) materials, Class 3, nondetonable; and water reactive materials, Class 3. NFPA 1, 2003 ed.

High Hazard Level 3 Contents. Materials that readily support combustion or present a physical hazard including, but not limited to: aerosols, Level 2 and Level 3; Class I, II, or III-A flammable or combustible liquids that are used or stored in normally closed containers or systems pressurized at less that 15 pounds per square inch (103.4kPa) gauge; consumer fireworks, 1.4G; flammable solids, other than dusts classified as High Hazard Level 2, stored, used, or generated in a manner creating a high fire hazard; organic peroxides, Class II and Class III; oxidizers, Class 2; Oxidizers, Class 3 that are used or stored in normally closed containers or systems pressurized at less than 15 pounds per square inch (103.4 kPa) gauge; oxidizing gases and oxidizing cryogenic liquids; unstable (reactive) materials, Class 2; and water-reactive materials, Class 2. NFPA 1, 2003 ed.

High Hazard Level 4 Contents. Materials that are acute health hazards including, but not limited to: corrosives, highly toxic materials, and toxic materials. NFPA 1, 2003 ed.

High Hazard Level 5 Contents. Hazardous production materials (HPM) used in the fabrication of semiconductors or semiconductor research and development. NFPA 1, 2003 ed.

High Hazard Materials. Materials that are combustible or flammable liquids, flammable gases, and combustible dusts. NFPA 221, 2000 ed.

High Hazard Occupancy. Areas that have high hazard buildings, materials, processes, or contents. NFPA 1710, 2004 ed.

High Temperature-Protective Clothing. Protective clothing designed to protect the wearer for short-term high temperature exposures. (SEE FIGURE H.7.) NFPA 472, 2002 ed.

Figure H.7 High Temperature Protective Clothing.

High-Energy Foam System. A device or system that adds the energy of a pressurized air source to the energy of a pressurized water source to create foam. NFPA 1145, 2000 ed.

High-Expansion Foam. Foams with expansion ratios ranging from 200:1 to approximately 1000:1. NFPA 1145, 2000 ed.

High-Idle Speed Control. A control or switch system that provides a means to increase the engine operating speed from an idle condition to a higher preset operating speed. NFPA 1901, 2003 ed.

High-Order Explosion. A rapid pressure rise or high-force explosion characterized by a shattering effect on the confining structure or container and long missile distances. NFPA 921, 2004 ed.

High-Point Anchor. A point above the trench used for attachment of rescue systems. NFPA 1006, 2003 ed.

High-Rise Building. A building greater than 23 m (75 ft) in height where the building height is measured from the lowest level of fire department vehicle access to the floor of the highest story that can be occupied. NFPA 5000, 2002 ed.

High-Risk Fuel. Class-IA, -IB, -IC, or -II liquids or Class-IIIA or -IIIB liquids heated to within 10°C (50°F) of their flash point, or pressurized to 174.4 kPa (25.3 psi) or more. NFPA 301, 2001 ed.

Highest Pulling Force (HPF). The pulling force that is achieved by the powered rescue tool while operating at the rated system input at the position of the arms or piston where the tool generates its greatest amount of force. NFPA 1936, 2005 ed.

Highest Spreading Force (HSF). The spreading force that is achieved by the powered rescue tool while operating at the rated system input at the position of the arms or piston where the tool generates its greatest amount of force. NFPA 1936, 2005 ed.

Highline System. A system of using rope suspended between two points for movement of persons or equipment over an area that is a barrier to the rescue operation, including systems capable of movement between points of equal or unequal height. NFPA 1670, 1999 ed.

Highly Toxic Gas. A chemical that has a median lethal concentration (LC50) in air of 200 parts per million by volume or less of gas or vapor, or 2 milligrams per liter or less of mist, fume, or dust, when administered by continuous inhalation for 1 hour (or less if death occurs within 1 hour) to albino rats weighing between 200 g and 300 g (0.44 lb and 0.66 lb) each. NFPA 1, 2003 ed.

Highly Volatile Liquid. A liquid with a boiling point of less than 20°C (68°F). NFPA 1, 2003 ed.

Highway. Any paved facility on which motor vehicles travel. NFPA 502, 2004 ed.

Hinge Pins. Pins that are used at either the swivel or point of articulation of an aerial device. NFPA 1914, 2002 ed.

Hitch. A knot that attaches to or wraps around an object so that when the object is removed, the knot will fall apart. (See Figure H.8.) NFPA 1670, 2004 ed.

HIV. Human immunodeficiency virus. NFPA 1581, 2005 ed.

Figure H.8 Hitch.

Hogging. Straining of the ship that tends to make the bow and stern lower than the middle portion. NFPA 1405, 2001 ed.

Holding Area. Location where the apparently uninjured aircraft occupants are transported. NFPA 424, 2002 ed.

Holding Valve. A one-way valve that maintains hydraulic pressure in a cylinder until it is activated to release. NFPA 1914, 2002 ed.

Hood. The interface component element of the protective ensemble designed to provide limited protection to the coat/helmet/SCBA facepiece interface area. (See Figure H.9.) NFPA 1976, 2000 ed.

Horizontal Center Plane. The plane that passes through the helmet and whose intersection with the helmet surface is equidistant from the top of the helmet at all points. NFPA 1971, 2000 ed.

Horizontal Exit. A way of passage from one building to an area of refuge in another building on approximately the same level, or a way of passage through or around a fire barrier to an area of refuge on approximately the same level in the same building that affords safety from fire and smoke originating from the area of incidence and areas communicating therewith. NFPA 101, 2003 ed.

Horizontal Stabilizer. That portion of an aircraft's structure that contains the elevators. NFPA 402, 2002 ed.

Hose Assembly. Hose with couplings attached to both ends. NFPA 1961, 2002 ed.

Hose Nozzle Valve. The terminal output end of a dispensing system hose. NFPA 1192, 2005 ed.

Hose Size. An expression of the internal diameter of the hose. NFPA 1961, 2002 ed.

Figure H.9 Protective Hood.

Hospital. A building or portion thereof used on a 24-hour basis for the medical, psychiatric, obstetrical, or surgical care of four or more inpatients. NFPA 101, 2003 ed.

Hot Brakes. A condition in which the aircraft's brake and wheel components have become overheated, usually due to excessive braking during landing. NFPA 402, 2002 ed.

Hot Spot. A particularly active part of a wildland fire. NFPA 1051, 2002 ed.

Hot Zone. The area immediately surrounding the physical location of a fire having a boundary that extends far enough from the fire to protect industrial fire brigade members positioned outside the hot zone from being directly exposed to flames, dense smoke, or extreme temperatures. NFPA 600, 2005 ed.

Hotel. A building or groups of buildings under the same management in which there are sleeping accommodations for more than 16 persons and primarily used by transients for lodging with or without meals. NFPA 101, 2003 ed.

House. A superstructure above the main deck. NFPA 1405, 2001 ed.

Hull Potential Monitor. A DC meter, analog or digital, that measures the potential of a metallic hull or metallic hull fittings as compared to a reference electrode. NFPA 1925, 2004 ed.

Human Exposure. Potential for injury or death to humans. NFPA 901, 2001 ed.

HVAC. Heating, ventilation, and air conditioning systems and their related components. NFPA 1620, 2003 ed.

Hydration. A fluid valance between water lost by normal functioning and oral intake of fluids in the form of liquid and foods that contain water. NFPA 1584, 2003 ed.

Hydrology. Effect of water, its movement and mechanics, in relation to bodies of water. NFPA 1006, 2003 ed.

Hyperbaric. Facility, building, or structure used to house chambers and all auxiliary service equipment for medical applications and procedures at pressures above normal atmospheric pressures. NFPA 99, 2005 ed.

Hypobaric. Facility, building, or structure used to house chambers and all auxiliary service equipment for medical applications and procedures at pressures below atmospheric pressures. NFPA 99, 2005 ed.

Hypergolic Material. Any substance that will spontaneously ignite or explode upon exposure to an oxidizer. NFPA 921, 2004 ed.

FIGURE I.1 Ignition.

Identical Rescue Tools. Powered rescue tools that are produced to the same engineering and manufacturing specifications. NFPA 1936, 2005 ed.

Identical SCBA. SCBA that are produced to the same engineering and manufacturing specifications. NFPA 1981, 2002 ed.

Identify. To select or indicate verbally or in writing using standard terms to establish the identity of; the fact of being the same as the one described. NFPA 472, 2002 ed.

Ignitable Liquid. Any liquid or the liquid phase of any material that is capable of fueling a fire, including a flammable liquid, combustible liquid, or any other material that can be liquefied and burned. NFPA 921, 2004 ed.

Ignitable Mixture. A gas-air, vapor-air, mist-air, or dust-air mixture, or combinations of such mixtures, that can be ignited by a sufficiently strong source of energy, such as a static electric discharge. NFPA 77, 2000 ed.

Ignition. The process of initiating self-sustained combustion. (See Figure I.1.) NFPA 921, 2004 ed.

Ignition Energy. The quantity of heat energy that should be absorbed by a substance to ignite and burn. (See Figure I.2.) NFPA 921, 2004 ed.

Ignition Sensitivity. A measure of the ease by which a cloud of combustible dust could be ignited. NFPA 499, 2004 ed.

Ignition Temperature. Minimum temperature a substance should attain in order to ignite under specific test conditions. (See Figure I.3.) NFPA 921, 2004 ed.

Ignition Time. The time between the application of an ignition source to a material and the onset of self-sustained combustion. NFPA 921, 2004 ed.

Immediately Dangerous to Life or Health (IDLH). Any condition that would pose an immediate or delayed threat to life, cause irreversible adverse health effects, or interfere with an individual's ability to escape unaided from a hazardous environment. NFPA 1670, 2004 ed.

Imminent Danger. A condition or practice in an occupancy or structure that poses a danger that could reasonably be expected to cause death, serious physical harm, or serious property loss. NFPA 1, 2003 ed.

Imminent Hazard. An act or condition that is judged to present a danger to persons or property that is so urgent and severe that it requires immediate corrective or preventive action. NFPA 1521, 2002 ed.

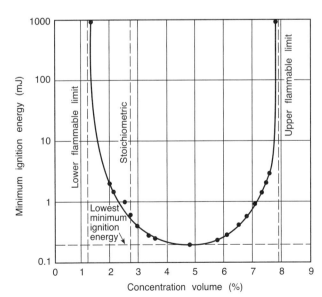

FIGURE I.2 Ignition Energy.

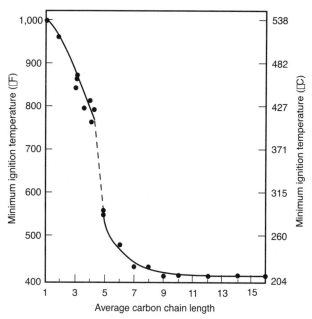

Figure I.3 Ignition Temperature.

Immunization. The process or procedure by which a person is rendered immune. NFPA 1581, 2005 ed.

Impact Analysis (Business Impact Analysis, BIA). A management level analysis that identifies the impacts of losing the entity's resources. *Note: The analysis measures the effect of resource loss and escalating losses over time in order to provide the entity with reliable data upon which to base decisions concerning hazard mitigation, recovery strategies, and continuity planning.* NFPA 1600, 2004 ed.

Impact Load. Sudden application of a force, which causes kinetic energy and momentum to be converted into other forms of energy. NFPA 1983, 2001 ed.

Important Building. A building that is considered not expendable in an exposure fire. NFPA 30, 2003 ed.

Impressed Current System. A cathodic protection system that uses an external power source to induce a DC electric current through an electrode (anode) that suppresses galvanic corrosion of the craft's hull. *Note: Typical external power sources are batteries, alternators, and rectified output from alternating current generators.* NFPA 1925, 2004 ed.

Improved Property. A piece of land or real estate upon which a structure has been placed, a marketable crop is growing (including timber), or other property improvement has been made. NFPA 1144, 2002 ed.

In Service. *As applied to fire hose:* The status of hose stored in a hose house, on a rack or reel, or on a fire apparatus that is available and ready for immediate use at an incident. NFPA 1962, 2003 ed. *As applied to ground ladders:* The status of a fire department ground ladder that has been inspected, maintained, and tested and currently is in use or available for use. NFPA 1932, 2004 ed. *As applied to SCBA:* Ready for immediate use. NFPA 1852, 2002 ed.

In Storage. The status of hose not readily available for use because it is not at the scene of an incident and not loaded on a vehicle that can transport it to the scene. NFPA 1962, 2003 ed.

In Use. The status of hose that has actually been deployed at an incident or during training whether or not water is running through the hose. NFPA 1962, 2003 ed.

In-Service Weight. The maximum actual vehicle weight under any conditions of mobile operation, sometimes referred to as gross vehicle weight. NFPA 1901, 2003 ed.

Incident. An occurrence, either human-caused or a natural phenomenon, that requires action or support by emergency services personnel to prevent or minimize loss of life or damage to property and/or natural resources. NFPA 1143, 2003 ed.

Incident Action Plan. The objectives reflecting the overall incident strategy, tactics, risk management, and member safety that are developed by the incident commander. Note: Incident action plans are updated throughout the incident. NFPA 1500, 2002 ed.

Incident Casualty. A person who is injured or killed as a result of responding to or handling an incident or who is the reason for the incident. NFPA 901, 2001 ed.

Incident Command System (ICS). The combination of facilities, equipment, personnel, procedures, and communications operating within a common organizational structure that has responsibility for the management of assigned resources to effectively accomplish stated objectives pertaining to an incident (as described in the document Incident Command System) or training exercise. NFPA 1670, 1999 ed.

Incident Commander. The person who is responsible for all decisions relating to the management of the incident and is in charge of the incident site. NFPA 472, 2002 ed.

Incident Location. The address or other identifiable area of an event. NFPA 450, 2004 ed.

Incident Management System. In disaster/emergency management applications, the combination of facilities, equipment, personnel, procedures, and communications operating within a common organizational structure with responsibility for the management of assigned resources to effectively accomplish stated objectives pertaining to an incident. (See Figure I.4.) NFPA 1600, 2004 ed.

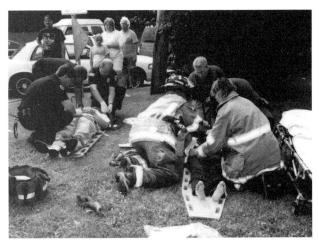

Figure I.4 Incident Management System.

Figure I.5 Incident Scene Rehabilitation.

Incident Record. The official file on an incident. NFPA 901, 2001 ed.

Incident Report. A document prepared by fire department personnel on a particular incident. NFPA 901, 2001 ed.

Incident Response Plan. Written procedures, including standard operating guidelines, for managing an emergency response and operation. NFPA 1670, 2004 ed.

Incident Safety Officer. An individual appointed to respond to or assigned at an incident scene by the incident commander to perform the duties and responsibilities specified in this standard. *Note: This individual can be the health and safety officer or it can be a separate function.* NFPA 1521, 2002 ed.

Incident Safety Plan. The strategies and tactics developed by the incident safety officer based upon the incident commander's incident action plan and the type of incident encountered. NFPA 1521, 2002 ed.

Incident Scene. The location where activities related to a specific incident are conducted. NFPA 1561, 2005 ed.

Incident Scene Rehabilitation. The tactical level management unit that provides for medical evaluation, treatment, monitoring, fluid and food replenishment, mental rest, and relief from climatic conditions of the incident. (See Figure I.5.) NFPA 1521, 2002 ed.

Incident Termination. The conclusion of emergency service operations at the scene of an incident, usually the departure of the last unit from the scene. NFPA 1561, 2005 ed.

Incidental Liquid Use or Storage. Use or storage as a subordinate activity to that which establishes the occupancy or area classification. NFPA 30, 2003 ed.

Incipient Fire Fighting. Fire fighting performed inside or outside of an enclosed structure or building when the fire has not progressed beyond incipient stage. NFPA 1081, 2001 ed.

Incipient Stage. Refers to the severity of a fire where the progression is in the early stage and has not developed beyond that which can be extinguished using portable fire extinguishers or handlines flowing up to 473 L/min (125 gpm). (See Figure I.6.) *Note: A fire is considered to be beyond the incipient stage when the use of thermal protective clothing or self-contained breathing apparatus is required or an industrial fire brigade member is required to crawl on the ground or floor to stay below smoke and heat.* NFPA 1081, 2001 ed.

Incline Plane. A lifting method that provides mechanical advantage by distributing the work required to lift a load over a distance along an incline rather than straight up and down. NFPA 1006, 2003 ed.

Figure I.6 Incipient Stage Fire.

Inclining Test. A test to determine the vessel displacement and center of gravity. NFPA 1925, 2004 ed.

Incompatible Material. Materials that, when in contact with each other, have the potential to react in a manner that generates heat, fumes, gases, or byproducts that are hazardous to life or property. NFPA 5000, 2002 ed.

Indicating Valve. A valve that has components that show if the valve is open or closed. *Note: Examples of indicating valves are outside screw and yoke (OS&Y) gate valves (*See Figure I.7*) and underground gate valves with indicator posts.* (See Figure I.8.) NFPA 1, 2003 ed.

Indirect Attack. Fire fighting operations involving the application of extinguishing agents to reduce the buildup of heat released from a fire without applying the agent directly onto the burning fuel. (See Figure I.9.) NFPA 1145, 2000 ed.

Indirect Medical Oversight. The administrative medical direction that can be in the form of system design, protocols and procedures, training, and quality assessment. NFPA 450, 2004 ed.

Individual Area of Specialization. The qualifications or functions of a specific job(s) associated with chemicals and/or containers used within an organization. NFPA 472, 2002 ed.

Indoor Area. An area that is within a building or structure having overhead cover, other than a structure qualifying as "weather protection." NFPA 55, 2005 ed.

Figure I.8 Post Indicator Valve.

Inductive Reasoning. The process by which a person starts from a particular experience and proceeds to generalizations. NFPA 921, 2004 ed.

Industrial Fire Brigade. An organized group of employees within an industrial occupancy who are knowledgeable, trained, and skilled in at least basic fire fighting operations, and whose full-time occupation might or might not be the provision of fire suppression and related activities for their employer. NFPA 600, 2005 ed.

Figure I.7 Outside Screw & Yoke Valve.

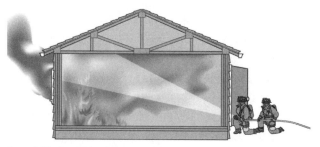

Figure I.9 Indirect Attack.

Industrial Fire Brigade Apparatus. An industrial fire brigade emergency response vehicle designed and intended primarily for fire suppression, rescue, or other specialized function that includes pumpers, foam apparatus, aerial ladders, rescue vehicles, and other such apparatus. NFPA 600, 2005 ed.

Industrial Fire Brigade Leader. An individual responsible for overseeing the performance or activity of other members. NFPA 1081, 2001 ed.

Industrial Fire Brigade Management. The individual designated by top management to be responsible for the organization, management, and functions of the industrial fire brigade. NFPA 600, 2005 ed.

Industrial Fire Brigade Training Coordinator. The designated company representative with responsibility for coordinating effective, consistent, and quality training within the industrial fire brigade training and education program. NFPA 600, 2005 ed.; NFPA 1081, 2001 ed.

Industrial Fire Department. An organization providing rescue, fire suppression, and related activities as well as emergency medical services, hazardous material operations, or other activities that occur at a single facility or facilities under the same management. NFPA 1500, 2002 ed.

Industrial Occupancy. An occupancy in which products are manufactured or in which processing, assembling, mixing, packaging, finishing, decorating, or repair operations are conducted. NFPA 5000, 2002 ed.

Industrial Packaging. Packaging grouped into three categories based on the strength of the packaging. NFPA 472, 2002 ed.

Industrial Supply Pump. A water pump mounted on a mobile foam fire apparatus with a rated capacity of 3000 gpm (12,000 L/min) or greater at 100 psi (700 kPa) net pump pressure. NFPA 1901, 2003 ed.

Industrialized Unit. A factory-built structure, designed for either permanent site installation or as a portable unit, and constructed to the requirements of a model building code or other state construction regulations. NFPA 901, 2001 ed.

Infection. The state or condition in which the body or a part of it is invaded by a pathogenic agent (microorganism or virus) that, under favorable conditions, multiplies and produces effects that are injurious. NFPA 1581, 2005 ed.

Infection Control Officer. The person or persons within the fire department who are responsible for managing the department infection control program and for coordinating efforts surrounding the investigation of an exposure. NFPA 1581, 2005 ed.

Infection Control Program. The fire department's formal policy and implementation of procedures relating to the control of infectious and communicable disease hazards where employees, patients, or the general public could be exposed to blood, body fluids, or other potentially infectious materials in the fire department work environment. NFPA 1500, 2002 ed.

Infectious Disease. An illness or disease resulting from invasion of a host by disease-producing organisms such as bacteria, viruses, fungi, or parasites. NFPA 1500, 2002 ed.

Inflatable Boat (IB). Any boat that achieves and maintains its intended shape and buoyancy through the medium of inflation. NFPA 1925, 2004 ed.

Information Officer. The individual who provides timely information to the media and others as authorized by the incident commander and functions as part of the command staff. NFPA 1561, 2005 ed.

Inherent Flame Resistance. As applied to textiles, flame resistance that is derived from an essential characteristic of the fiber or polymer from which the textile is made. NFPA 1971, 2000 ed.

Initial Attack. An aggressive suppression action consistent with fire fighter and public safety and values to be protected. NFPA 1051, 2002 ed.

Initial Attack Apparatus. Fire apparatus with a permanently mounted fire pump of at least 250 gpm (1000 L/min) capacity, water tank, and hose body whose primary purpose is to initiate a fire suppression attack on structural, vehicular, or vegetation fires, and to support associated fire department operations. NFPA 1901, 2003 ed.

Initial Attack Line. The first hose stream placed in service by a company at the scene of a fire in order to protect lives or to prevent further extension of fire while additional lines are laid and placed in position. NFPA 1410, 2005 ed.

Initial Full Alarm Assignment. Those personnel, equipment, and resources ordinarily dispatched upon notification of a structural fire. NFPA 1710, 2004 ed.

Initial Rapid Intervention Crew (IRIC). Two members of the initial attack crew who are assigned for rapid deployment to rescue lost or trapped members. NFPA 1710, 2004 ed.

Initiating Device Circuit. A circuit to which automatic or manual initiating devices are connected where the signal received does not identify the individual device operated. NFPA 72, 2002 ed.

Initiative. A fire or life safety program that targets a specific issue and audience(s) and is terminated when program goals are achieved. NFPA 1035, 2005 ed.

Injector. A device used in a discharge or intake line to force foam concentrate into the water stream. NFPA 1145, 2000 ed.

Injury. Physical damage to a person suffered as the result of an incident that requires (or should require) treatment by a practitioner of medicine, a registered EMT, or a paramedic within one year of the incident (regardless of whether treatment was actually received) or that results in at least one day of restricted activity immediately following the incident. NFPA 901, 2001 ed.

Inner Perimeter. The area that is secured to allow effective command, communication, and coordination control and to allow for safe operations to deal with an emergency, including the immediate ingress and egress needs of emergency response personnel and vehicles. NFPA 424, 2002 ed.

Inseam Length. Lower torso garment measurement along inseam from crotch seam to bottom edge of cuff. NFPA 1977, 2005 ed.

Inside Liquid Storage Area. A room or building used for the storage of liquids in containers or portable tanks, separated from other types of occupancies. NFPA 30, 2003 ed.

Inside Room. A room totally enclosed within a building and having no exterior walls. NFPA 30, 2003 ed.

Insole. The inner part of the protective footwear upon which the foot rests and that conforms to the bottom of the foot. NFPA 1971, 2000 ed.

Inspect. To determine the condition or operation of a component(s) by comparing its physical, mechanical, and/or electrical characteristics with established standards, recommendations, and requirements through examination by sight, sound, or feel. NFPA 1915, 2000 ed.

Instability. A condition of a mobile unit in which the sum of the moments tending to overturn the unit exceeds the sum of the moments tending to resist overturning. NFPA 1901, 2003 ed.

Instant Recall Recorder. A device that records voice conversations and that is intended to provide a telecommunicator with a means to review such conversations in real time. NFPA 1221, 2002 ed.

Instructor. *As applied to live fire training:* An individual qualified by the authority having jurisdiction to deliver fire fighter training, who has the training and experience to supervise students during live fire training evolutions. NFPA 1403, 2002 ed. *As applied to live fire service vehicle training:* An individual deemed qualified by the authority having jurisdiction to deliver training in the operation of fire service vehicles. NFPA 1451, 2002 ed.

Instructor-in-Charge. An individual qualified as an instructor and designated by the authority having jurisdiction to be in charge of the live fire training evolution. NFPA 1403, 2002 ed.

Insurance. Transfer by contract of funds (premium) in exchange for payment on losses that might occur. NFPA 1250, 2004 ed.

Intake. The process of collecting the comprehensive background information for the juvenile and family regarding the incident(s) that brought the juvenile to the program. NFPA 1035, 2005 ed.

Intake Connection Size. The nominal size of the first fire hose connection from the pump on an intake. NFPA 1901, 2003 ed.

Intake Pressure. The pressure on the intake passageway of the pump at the point of gauge attachment. NFPA 1911, 2002 ed.

Intake Relief Valve. A relief valve piped to the intake manifold of a pump and designed to automatically relieve excessive pressure from the incoming flow of water by discharging water to the environment. NFPA 1901, 2003 ed.

Integrity. The ability of an ensemble or element to remain intact and provide continued minimum performance. NFPA 1851, 2001 ed.

Intelligent Transportation System. A means of electronic communications or information processing used singly or in combination to improve the efficiency or safety of a surface transportation system. NFPA 1221, 2002 ed.

Interagency Network. A group of agencies (public safety, social services, education, mental health, health care providers, law enforcement, and juvenile justice) working in a formal/informal partnership to address juvenile firesetting. NFPA 1035, 2005 ed.

Interface Area. An area of the body where the protective garments, helmet, gloves, footwear, or SCBA facepiece meet (i.e., the protective coat/helmet/SCBA facepiece area, protective coat/protective trouser area, the protective coat/glove area, and the protective trouser/footwear area). NFPA 1971, 2000 ed.

Interface Components. Elements of the proximity protective ensemble that are designed to provide limited protection to interface areas. NFPA 1971, 2000 ed.

Intergranular Corrosion. A corrosive attack on metal at the grain boundary. NFPA 1150, 2004 ed.

Interior. A sheltered location not exposed to the environment. NFPA 1901, 2003 ed.

Interior Structural Fire Fighting. The physical activity of fire suppression, rescue, or both, inside of buildings or enclosed structures that are involved in a fire beyond the incipient stage. NFPA 600, 2005 ed.

Interlining. Any textile that is incorporated into any garment as a layer between the outer and inner layers. NFPA 1977, 2005 ed.

Intermediate Level of Supervision. A level of supervision within the incident management system that groups fire companies and other resources working toward common objectives or in a particular area under a supervisor responsible for the objective(s) or area. NFPA 1561, 2005 ed.

International Air Transport Association (IATA). An international group composed of the major airlines of the world that reviews aviation policy including safety items. NFPA 402, 2002 ed.

International Civil Aviation Organization (ICAO). An international aviation body, operating under the auspices of the United Nations, that produces technical safety documents for civil air transport. NFPA 402, 2002 ed.

International Shore Connection. A universal connection to the vessel's fire main to which a shoreside fire fighting water supply can be connected. NFPA 1405, 2001 ed.

Intersecting Trench. A trench where multiple trench cuts or legs converge at a single point. NFPA 1006, 2003 ed.

Intervention. A formal response to firesetting behavior that includes assessment, education, referral, counseling, medical services, social services, and juvenile justice sanctions. NFPA 1035, 2005 ed.

Interview. The process of meeting with the juvenile firesetter and the family to determine the severity of the problem. NFPA 1035, 2005 ed.

Investigation. A systematic inquiry or examination. NFPA 1033, 2003 ed.

Investigator's Special Tools. Tools of a specialized or unique nature that might not be required for every fire investigation. NFPA 1033, 2003 ed.

Ironing. A term used for the damage caused to the bottom of a base rail by misalignment or malfunction of the rollers, which causes wear or indentation of the base rail material. NFPA 1914, 2002 ed.

Irritant Gas. A chemical that is not corrosive, but that causes a reversible inflammatory effect on living tissue by chemical action at the site of contact. *Note: A chemical is a skin irritant if, when tested on the intact skin of albino rabbits by the methods of 16 CFR 1500.41, for an exposure of 4 or more hours or by other appropriate techniques, it results in an empirical score of 5 or more. A chemical is classified as an eye irritant if so determined under the procedure listed in 16 CFR 1500.42, or other appropriate techniques.* NFPA 55, 2005 ed.

Isochar. A line on a diagram connecting points of equal char depth. NFPA 921, 2004 ed.

Isolated Storage. Storage in a different storage room, or in a separate and detached building located at a safe distance. NFPA 1, 2003 ed.

Isolation. The process by which an area is rendered safe through mitigation of dangerous energy forms. NFPA 1006, 2003 ed.

Isolation System (or Isolation Devices). An arrangement of devices, applied with specific techniques, that collectively serve to isolate a victim of a trench or excavation emergency from the surrounding product (e.g., soil, gravel, or sand). NFPA 1670, 2004 ed.

J

Jacob's Ladder. A rope or chain ladder with rigid rungs. NFPA 1405, 2001 ed.

Jet Blast. The thrust-producing exhaust from a jet engine. NFPA 402, 2002 ed.

Jet Drive. A vessel propelled by reaction to a water stream. NFPA 1925, 2004 ed.

Job. An organized segment of instruction designed to develop sensory motor skills or technical knowledge. NFPA 1071, 2000 ed.

Job Performance Requirement. A statement that describes a specific job task, lists the items necessary to complete the task, and defines measurable or observable outcomes and evaluation areas for the specific task. NFPA 1000, 2000 ed.

Joint Aviation Authority (JAA). An agency in Europe charged with the responsibility of regulating safety in civil aviation. NFPA 402, 2002 ed.

Joule. The preferred SI unit of heat, energy, or work; there are 4.184 joules in a calorie, and 1055 joules in a British thermal unit (Btu). *Note: A watt is a joule/second.* NFPA 921, 2004 ed.

Jurisdiction. Any governmental unit or political division or subdivision including, but not limited to, township, village, borough, parish, city, county, state, commonwealth, province, freehold, district, or territory over which the governmental unit exercises power and authority. NFPA 1141, 2003 ed.

Juvenile Firesetter. A person, through the age of 18, or as defined by the authority having jurisdiction, who is involved in the act of firesetting. NFPA 1035, 2005 ed.

Juvenile Firesetter Intervention Specialist I. The individual who has demonstrated the ability to conduct an intake/interview with a firesetter and his or her family using prepared forms and guidelines and who, based on program policies and procedures, determines the need for referral for counseling and/or implements educational intervention strategies to mitigate effects of firesetting behavior. NFPA 1035, 2005 ed.

Juvenile Firesetter Intervention Specialist II. The individual who has demonstrated the ability to manage juvenile firesetting intervention program activities and the activities of Juvenile Firesetter Intervention Specialist I. NFPA 1035, 2005 ed.

Keel. The principal structural member of a ship, running fore and aft on the centerline, extending from bow to stern, forming the backbone of the vessel to which the frames are attached. NFPA 1405, 2001 ed.

Keying. Activating the transmitter by pressing the push-to-talk button. NFPA 1221, 2002 ed.

Kilowatt. A measurement of energy release rate. NFPA 921, 2004 ed.

Kink Test Pressure. A pressure equal to at least 1.5 times the service test pressure. NFPA 1961, 2002 ed.

Kitchen. An area designated for storage, preparation, cooking, and serving of food for members. NFPA 1581, 2005 ed.

Knee Circumference. Lower torso garment measurement 355 mm (14 in.) below crotch seam, from folded edge to folded edge, and multiplied by 2 to obtain circumference. NFPA 1977, 2005 ed.

Knot. A fastening made by tying together lengths of rope or webbing in a prescribed way. NFPA 1670, 2004 ed.

Knot Wood Irregularity. A portion of a branch or limb embedded in the tree and cut during the process of lumber manufacture. NFPA 1931, 2004 ed.

Knuckle. A point of connection between the upper and lower booms of an articulating device; the point at which lower and upper booms are hinged together. NFPA 1901, 2003 ed.

kPa. Kilopascals. NFPA 1410, 2005 ed.

kPag. Gauge pressure in Kilopascals. NFPA 58, 2004 ed.

L

L/min. Liters per minute. NFPA 1410, 2005 ed.

Label. A visual indication whether in pictorial or word format that provides for the identification of a control, switch, indicator, gauge, or the display of information useful to the operator. NFPA 1906, 2001 ed.

Ladder. A device consisting of two beams (side rails) joined at regular intervals by crosspieces called rungs on which a person is supported during climbs for ascending or descending. (See Figure L.1.) NFPA 1931, 2004 ed.

Ladder Cradle. A structural component that supports an aerial ladder when it is bedded. NFPA 1914, 2002 ed.

Ladder Locks. The mechanical locks or pawls that prevent movement of the sections of an aerial device when the power is shut off or in the event of loss of pressure in hydraulic circuits. NFPA 1914, 2002 ed.

Ladder Nesting. The procedure whereby ladders of different sizes are positioned partially within one another to reduce the amount of space required for their storage on the apparatus. (See Figure L.2.) NFPA 1931, 2004 ed.

Ladder Pipe. A monitor that is fed by a hose and that holds and directs a nozzle while attached to the rungs of a vehicle-mounted aerial ladder. NFPA 1965, 2003 ed.

Ladder Section. A structural member normally of an open "U" truss-type design that includes the rungs and comprises the base or fly section of an aerial ladder. NFPA 1901, 2003 ed.

Ladder Shank. Reinforcement to the shank area of footwear designed to provide additional support to the instep when standing on a ladder rung. NFPA 1971, 2000 ed.

Large Diameter Hose. A hose of $3\frac{1}{2}$ in. (90 mm) size or larger. NFPA 1142, 2001 ed.; NFPA 1961, 2002 ed.

Large Stream Device. Any device that discharges water at a flow rate greater than 400 gpm (1514 L/min). NFPA 1963, 1998 ed.

Laser Target. A square or rectangular plastic device used in conjunction with a laser instrument to set the line and grade of pipe. NFPA 1670, 2004 ed.

Layering. The systematic process of removing debris from the top down and observing the relative location of artifacts at the fire scene. NFPA 921, 2004 ed.

LC50 (Lethal Concentration50). The concentration of agent in water, usually expressed as milligrams of agent in a liter or solution, that results in the death of 50 percent of the aquatic test specimens within a specified time frame. NFPA 1150, 2004 ed.

Figure L.1 Ladder.

Figure L.2 Ladder Nesting.

LD50 (Lethal Dosage50). The dosage of a chemical, usually expressed as milligrams of the chemical per kilogram of body weight of the test animal, at which 50 percent of the test animals die within a specified time frame. NFPA 1150, 2004 ed.

Leader Line. A hose line supplying one or more smaller lines, as in a wyed line. NFPA 1410, 2005 ed.

Leak. A continuous stream of liquid escaping from a hose, pipe, coupling, connection, or other confining structure at any point where the escape should not occur. NFPA 1914, 2002 ed.

Lesson. A component of a program in which the educator directly presents fire or life safety information to a group. NFPA 1035, 2005 ed.

Level I Refurbishing. The assembly of a new fire apparatus by the use of a new chassis frame, driving and crew compartment, front axle, steering and suspension components, and the use of either new components or components from an existing apparatus for the remainder of the apparatus. NFPA 1912, 2001 ed.

Level II (ASNT). A tested and experienced level of proficiency for a nondestructive testing technician. NFPA 1914, 2002 ed.

Level II Refurbishing. The upgrade of major components or systems of a fire apparatus with components or systems that comply with the applicable standards in effect at the time the original apparatus was manufactured. NFPA 1912, 2001 ed.

Leveling Linkages. The components and controls for achieving a level position of the platform. NFPA 1914, 2002 ed.

Lever-Type Control. A control in which the handle operates along the axis of the nozzle. NFPA 1964, 2003 ed.

Levers. Tools that have a relationship of load/fulcrum/force to create mechanical advantage and move a load. NFPA 1006, 2003 ed.

Liability. Legal responsibility and accountability for an act or process related to a program. NFPA 1035, 2005 ed.

Liaison Officer. A member of the command staff, responsible for coordinating with representatives from cooperating and assisting agencies. NFPA 1561, 2005 ed.

Life Rail, Deck Rail, or Lifeline. A single rail or the entire assembly of stanchions, lines, or rails, including hardware, gates, and so forth, surrounding weather decks and designed to reduce falls overboard. NFPA 1925, 2004 ed.

Life Safety Harness. A system component; an arrangement of materials secured about the body used to support a person during fire service rescue. (SEE FIGURES L.3 AND L.4.) NFPA 1006, 2003 ed.

FIGURE L.3 Seat Harness.

Life Safety Rope. A compact but flexible, torsionally balanced, continuous structure of fibers produced from strands that are twisted, plaited, or braided together and that serve primarily to support a load or transmit a force from the point of origin to the point of application. (SEE FIGURE L.5.) NFPA 1500, 2002 ed.

Life Safety Systems. Those systems that enhance or facilitate evacuation smoke control, compartmentalization, and/or isolation. NFPA 1031, 2003 ed.

FIGURE L.4 Chest Harness.

FIGURE L.5 Life Safety Rope.

Lift. *As applied to fire pump systems on fire apparatus:* The vertical height that water must be raised during a drafting operation, measured from the surface of a static source of water to the centerline of the pump intake. NFPA 1911, 2002 ed. *As applied to pre-incident planning:* A mechanically or electrically operated platform used to work at various heights within a building. NFPA 1620, 2003 ed.

Lifting Tools. Hydraulic, pneumatic, mechanical, or manual tools that can lift heavy loads. NFPA 1006, 2003 ed.

Light Frame Construction. Structures that have framework made out of wood or other lightweight materials. NFPA 1006, 2003 ed.

Light Use. A designation of system components or manufactured systems designed for light-use loads and escape. NFPA 1983, 2001 ed.

Lightweight Construction. The use of nonferrous metals or plastics or a reduction in weight by the use of advanced engineering practices resulting in a weight saving without sacrifice of strength or efficiency. NFPA 414, 2001 ed.

Likelihood. Frequency, probability, or their combination. NFPA 551, 2004 ed.

Limber Holes. Holes in hull framing members to permit draining of liquids. NFPA 1925, 2004 ed.

Limited Care Facility. A building or portion of a building used on a 24-hour basis for the housing of four or more persons who are incapable of self-preservation because of age; physical limitations due to accident or illness; or limitations such as mental retardation/developmental disability, mental illness, or chemical dependency. NFPA 99, 2005, ed.

Line. *As applied to emergency operations:* One or more lengths of connected fire hose. NFPA 1410, 2005 ed. *As applied to marine vessels:* Rope, when in use. NFPA 1925, 2004 ed.

Line Voltage Circuit, Equipment, or System. An ac or dc electrical circuit, equipment, or system where the voltage to ground or from line to line is 30 volts (V) rms (ac) or 42.4 V peak (dc) or greater. NFPA 1901, 2003 ed.

Line Voltage Conductor. An ungrounded current-carrying conductor of a line voltage circuit. NFPA 1901, 2003 ed.

Liner. An inner component of a helmet or garment designed to provide warmth. NFPA 1977, 2005 ed.

Liner System. The combination of the moisture barrier and thermal barrier as used in a garment. NFPA 1851, 2001 ed.

Lining. Any material that is permanently attached and used to cover or partially cover the inside surface area of a protective garment. NFPA 1977, 2005 ed.

Lip Collapse. A collapse of the trench lip, usually subsequent to surcharge loading, impact damage from the excavating bucket, and/or inherent cohesive properties of the soil type. NFPA 1006, 2003 ed.

Lip (Trench Lip). The area 2 ft horizontal and 2 ft vertical (0.61 m × 0.61 m) from the top edge of the trench face. NFPA 1006, 2003 ed.

Liquefied Compressed Gases. Gases that are contained in a packaging under the charged pressure and are partially liquid at a temperature of 20°C (68°F). NFPA 1, 2003 ed.

Liquefied Natural Gas (LNG). A fluid in the liquid state that is composed predominantly of methane and that can contain minor quantities of ethane, propane, nitrogen, or other components normally found in natural gas. NFPA 57, 2002 ed.

Liquefied Petroleum Gas (LP-Gas). Any material having a vapor pressure not exceeding that allowed for commercial propane that is composed predominantly of the following hydrocarbons, either by themselves or as mixtures: propane, propylene, butane (normal butane or isobutane), and butylenes. NFPA 58, 2004 ed.

Liquid Borne Pathogen. An infectious bacteria or virus carried in human, animal, or clinical body fluids, organs, or tissues. NFPA 1971, 2000 ed.

Liquid Penetrant Inspection. A nondestructive inspection method used to locate and determine the severity of surface discontinuities in materials, based on the ability of a liquid to penetrate into small openings, such as cracks. NFPA 1914, 2002 ed.

Liquid Splash-Protective Clothing. Multiple items of compliant protective clothing and equipment products that provide protection from some risks, but not all risks, of hazardous materials emergency incident operations involving liquids. (See Figure L.6.) NFPA 1992, 2005 ed.

Liquid Splash-Protective Footwear. The element of the protective ensemble or the item of protective clothing that provides liquid chemical protection and physical protection to the feet, ankles, and lower legs. NFPA 1992, 2005 ed.

FIGURE L.6 Liquid Splash-Protective Clothing.

Liquid Splash-Protective Garment. The element of the protective ensemble or the item of protective clothing that provides liquid chemical protection to the upper and lower torso, arms and legs, excluding the head, hands, and feet. NFPA 1992, 2005 ed.

Liquid Splash-Protective Glove. The element of the protective ensemble, or the item of protective clothing that provides liquid chemical protection to the hands and wrists. NFPA 1992, 2005 ed.

Liquid Surge. The force imposed upon a fire apparatus by the contents of a partially filled water or foam concentrate tank when the vehicle is accelerated, decelerated, or turned. NFPA 1002, 2003 ed.

Liquid Warehouse. A separate, detached building or attached building used for warehousing-type operations for liquids. NFPA 30, 2003 ed.

Liquidborne Pathogen. An infectious bacteria or virus carried in human, animal, or clinical body fluids, organs, or tissues. NFPA 1951, 2005 ed.

List. An inclination to one side; a tilt. NFPA 1405, 2001 ed.

Listed. Equipment, materials, or services included in a list published by an organization that is acceptable to the authority having jurisdiction and concerned with evaluation of products or services, that maintains periodic inspection of production of listed equipment or materials or periodic evaluation of services, and whose listing states that either the equipment, material, or service meets appropriate designated standards or has been tested and found suitable for a specified purpose. NFPA Official Definition.

Litter. A transfer device designed to support and protect a victim during movement. NFPA 1670, 2004 ed.

Litter Attendant. A person who both accompanies and physically manages the litter. NFPA 1670, 2004 ed.

Live Fire. Any unconfined open flame or device that can propagate fire to the building or other combustible materials. NFPA 1403, 2002 ed.

Live Load. Forces acting on the aerial device from personnel, portable equipment, water, and nozzle reaction. NFPA 1901, 2003 ed.

Load. That which is being lowered or raised by rope in a high angle system. NFPA 1670, 2004 ed.

Load Limit Indicator. A load indicator or an instruction plate, visible at the operator's position, that shows the recommended safe load at any condition of an aerial device's elevation and extension. NFPA 1901, 2003 ed.

Load Stabilization. The process of preventing a load from shifting in any direction. NFPA 1006, 2003 ed.

Load Test. A method of preloading a rope rescue system to ensure all components are set properly to sustain the expected load. NFPA 1670, 1999 ed.

Load-Bearing Connector. An auxiliary equipment system component; a device used to join other system components including but not limited to carabiners, rings, rapid links, and snap-links. NFPA 1983, 2001 ed.

Local Area. A geographic area that includes the defined response area and receiving facilities for an EMS agency. NFPA 473, 2002 ed.

Local Circuit. A circuit that does not depend on the receipt of alarms over box circuits or the retransmission of alarms over dispatch circuits. NFPA 1221, 2002 ed.

Local Emergency Response Plan. The plan promulgated by the authority having jurisdiction, such as the local emergency planning committee for the community or a facility. NFPA 472, 2002 ed.

Locating Devices. Devices utilized to locate victims in rescue incidents and structural components, including but not limited to voice, seismic, video, K-9, and fiber optic. NFPA 1006, 2003 ed.

Lock Box. A locked container often used to store building entry keys, plans, and related data. *Note: Keys to open these containers are assigned only to selected individuals, such as representatives of the local fire department or police department.* NFPA 1620, 2003 ed.

Lockout. A method for keeping equipment from being set in motion and endangering workers. (See Figure L.7.) NFPA 1670, 2004 ed.

Lodging or Rooming House. A building or portion thereof that does not qualify as a one- or two-family dwelling, which provides sleeping accommodations for a total of 16 or fewer people on a transient or permanent basis, without personal care services, with or without meals, but without separate cooking facilities for individual occupants. NFPA 101, 2003 ed.

Logging Voice Recorder. A device that records voice conversations and automatically logs the time and date of such conversations; normally, a multichannel device that keeps a semipermanent record of operations. NFPA 1221, 2002 ed.

Logistics. The incident management section responsible for providing facilities, services, and materials for the incident. NFPA 1143, 2003 ed.

Lookout. A person designated to observe the fire or a portion of a fire and warn the crew when there is a change in fire activity or when there is danger of becoming trapped. NFPA 1051, 2002 ed.

FIGURE L.7 Lockout.

Loop. An element of a knot created by forming a complete circle in a rope. (See Figure L.8.) NFPA 1006, 2003 ed.

Loose House. A separate detached building in which unbaled combustible fibers are stored. NFPA 1, 2003 ed.

Loss. The unintentional decline in or disappearance of value arising from an incident. NFPA 1250, 2004 ed.

FIGURE L.8 Loop.

FIGURE L.9 Low-Angle Operation.

Low Angle. Refers to an environment in which the load is predominately supported by itself and not the rope rescue system (e.g., flat land or mild sloping surface). (See Figure L.9.) NFPA 1670, 2004 ed.

Low Explosive. An explosive that has a reaction velocity of less than 1000 m/s (3000 ft/sec). NFPA 921, 2004 ed.

Low Hazard. Contents of such low combustibility that no self-propagating fire therein can occur. NFPA 520, 2005 ed.

Low Voltage Circuit, Equipment, or System. An electrical circuit, equipment, or system where the voltage does not exceed 30 volts (V) rms (ac) or 42.4 V peak (dc), usually 12 V dc in fire apparatus. NFPA 1901, 2003 ed.

Low-Band VHF. Radio frequencies of 30 MHz to 50 MHz. NFPA 1221, 2002 ed.

Low-Energy Foam System. A device or system that uses only energy produced by the velocity of the water stream to create foam. NFPA 1145, 2000 ed.

Low-Expansion Foam. Foams with expansion ratios up to 20:1. NFPA 1145, 2000 ed.

Low-Order Explosion. A slow rate of pressure rise or low-force explosion characterized by a pushing or dislodging effect on the confining structure or container and by short missile distances. NFPA 921, 2004 ed.

Low-Voltage Circuit, Equipment, or System. An electrical circuit, equipment, or system where the voltage does not exceed 30 volts (V) rms (ac) or 42.4 V peak (dc), usually 12 V dc in fire apparatus. NFPA 1901, 1999 ed.

Lower Flammable Limit. That concentration of a combustible material in air below which ignition will not occur. *Note: Lower flammable limit is also known as the lower explosive limit (LEL).* NFPA 329, 2005 ed.

Lower Torso. The area of body below the waist including the legs but excluding the ankles and feet. NFPA 1971, 2000 ed.

Lowering System. A rope rescue system used to lower a load under control. NFPA 1670, 2004 ed.

Lowest Pulling Force (LPF). The pulling force that is achieved by the powered rescue tool while operating at the rated system input at the position of the arms or piston where the tool generates its least amount of force. NFPA 1936, 2005 ed.

Lowest Spreading Force (LSF). The spreading force that is achieved by the powered rescue tool while operating at the rated system input at the position of the arms or piston where the tool generates its least amount of force. NFPA 1936, 2005 ed.

LP-Gas Container. A vessel, including cylinders, tanks, portable tanks, and cargo tanks, used for the transporting or storing of LP-Gases. NFPA 1, 2003 ed.

Magnesium. Refers to either pure metal or alloys having the generally recognized properties of magnesium marketed under different trade names and designations. NFPA 402, 2002 ed.

Magnetic Particle Inspection. A nondestructive inspection method used to locate discontinuities in ferromagnetic materials by magnetizing the material and then applying an iron powder to mark and interpret the patterns that form. NFPA 1914, 2002 ed.

Main Deck. The uppermost continuous deck of a ship that runs from bow to stern. NFPA 1405, 2001 ed.

Maintenance. *As applied to fire apparatus:* The act of servicing a fire apparatus or a component within the time frame prescribed by the authority having jurisdiction, based on manufacturer's recommendations, local experience, and operating conditions in order to keep the vehicle and its components in proper operating condition. NFPA 1915, 2000 ed. *As applied to protective ensembles:* Procedures for inspection, repair, and removal from service of protective ensembles or ensemble elements. NFPA 1994, 2001 ed.

Maintenance Kits. Items required for maintenance and inspection that include, but are not limited to, the following: manufacturer product specifications; preventive maintenance checklists; periodic logbook records; inventory equipment lists; appropriate fluids, parts, and hardware; and testing instruments as required. NFPA 1006, 2003 ed.

Major Conversion. A change in service of the vessel from some other use to use as a marine fire fighting vessel. NFPA 1925, 2004 ed.

Major Fire Hazard Area. Includes machinery spaces, engine casing, exhaust tunnels and equivalents, special category spaces, and any compartment where the proximity of combustible materials, flammable liquids, and potential sources of ignition can promote a fire. NFPA 1925, 2004 ed.

Major Repair Garage. A building or portions of a building where major repairs, such as engine overhauls, painting, body and fender work, and repairs that require draining of the motor vehicle fuel tank are performed on motor vehicles, including associated floor space used for offices, parking, or showrooms. NFPA 30A, 2003 ed.

Major Stress Seams. Classes of seams that designate minimum sewn seam requirements. NFPA 1975, 2004 ed.

Major Stress Seams Class I. The seat seams, side seams, and inseams of pants; the seat seams, side seams, inseams, and waist seams in the bottom portion of coveralls; and the yoke(s) seams, side seams, sleeve set and close seams, and shoulder seams for the upper portion of coveralls. NFPA 1975, 2004 ed.

Major Stress Seams Class II. The yoke(s) seams, side seams, sleeve set and close seams, and shoulder seams for knit fabrics and woven shirting fabrics. NFPA 1975, 2004 ed.

Management. The collective body of those who direct the operations of the organization. NFPA 1401, 2001 ed.

Manual Defibrillator. A device that delivers an electric shock through the chest wall to the heart and that requires operation by trained medical personnel. NFPA 450, 2004 ed.

Manual Fire Alarm Box. A manually operated device used to initiate an alarm signal. (SEE FIGURE M.1.) NFPA 72, 2002 ed.

Manufactured System. Preassembled system, sold as a unit by the manufacturer and tested as a complete assembly. NFPA 1983, 2001 ed.

FIGURE M.1 Fire Alarm Box.

Manufacturer. The person or persons, company, firm, corporation, partnership, or other organization responsible for turning raw materials or components into a finished product. NFPA 1901, 2003 ed.

Manufacturer's Lot. An identifiable series of products that can be the same as or a subset of a production lot; used by the manufacturer for quality control or identification purposes. NFPA 1983, 2001 ed.

Manufacturer's Recommendation (Specification). Any requirement or suggestion a fire apparatus builder or component producer makes in regard to care and maintenance of its product(s). NFPA 1915, 2000 ed.

Manufacturer's Specifications. Any requirement or service bulletin an emergency response vehicle builder or component producer provides with regard to the use, care, and maintenance of its product(s). NFPA 1071, 2000 ed.

Marine Motor Fuel Dispensing Facility. A motor fuel dispensing facility at or adjacent to shore, a pier, a wharf, or a floating dock where motor fuels are dispensed into the fuel tanks of marine craft. NFPA 30A, 2003 ed.

Marine Rescue and Fire Fighting. The fire fighting action taken to prevent, control, or extinguish fire involved in or adjacent to a marine vessel and the rescue actions for occupants using normal and emergency routes for egress. NFPA 1710, 2004 ed.

Marine Vessel. A water craft or other artificial contrivance used as a means of transportation in or on the water. NFPA 1, 2003 ed.

Marking Systems. Various systems used to mark hazards, victim location, and pertinent structural information. NFPA 1006, 2003 ed.

Mask. A device designed to limit exposure of the nasal, oral, respiratory, or mucosal membranes to airborne pathogens. NFPA 1581, 2005 ed.

Master. The captain of a merchant ship. NFPA 1405, 2001 ed.

Master Stream. A portable or fixed fire fighting appliance supplied by either hose lines or fixed piping and that has the capability of flowing in excess of 1140 L/min (300 gpm) of water or water-based extinguishing agent. (See Figure M.2.) NFPA 600, 2005 ed.

Master Stream Nozzle. A nozzle with a rated discharge of 1325 L/min (350 gpm) or greater. NFPA 1964, 2003 ed.

Match. To provide with a counterpart. NFPA 472, 2002 ed.

Mate. A deck officer on a merchant ship ranking below the master. NFPA 1405, 2001 ed.

FIGURE M.2. Master Stream Device.

Material First Ignited. The fuel that is first set on fire by the heat of ignition; to be meaningful, both a type of material and a form of material should be identified. NFPA 921, 2004 ed.

Material Safety Data Sheet (MSDS). A form, provided by manufacturers and compounders (blenders) of chemicals, containing information about chemical composition, physical and chemical properties, health and safety hazards, emergency response, and waste disposal of the material. (See Figure M.3.) NFPA 472, 2002 ed.

Maximum Extended Length. The total length of the extension ladder when all fly sections are fully extended and all pawls are engaged. NFPA 1931, 2004 ed.

Maximum Operating Pressure. The maximum pressure at which the device is designed to be operated. NFPA 1965, 2003 ed.

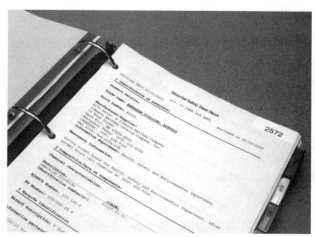

FIGURE M.3 MSDS.

Maximum Pump Close-Off Pressure. The maximum pump discharge pressure obtained with all discharge outlets closed, with the pump primed and running with the pump drive engine operating at maximum obtainable speed, and with the pump intake pressure at atmospheric pressure or less. NFPA 1901, 2003 ed.

Maximum Rated Pressure. The maximum pressure at which the manufacturer determines it is safe to operate the nozzle. NFPA 1964, 2003 ed.

Maximum Working Load. Weight supported by the life safety rope and system components that must not be exceeded. NFPA 1670, 2004 ed.

Means of Access. The method by which entry or approach is made by emergency apparatus to structures. *Note: Examples of means of access include roadways, fire lanes, and parking lots.* NFPA 1141, 2003 ed.

Means of Egress. A continuous and unobstructed way of travel from any point in a building or structure to a public way consisting of three separate and distinct parts: (1) the exit access, (2) the exit, and (3) the exit discharge. NFPA 101, 2003 ed.

Means of Escape. A way out of a building or structure that does not conform to the strict definition of means of egress but does provide an alternate way out. NFPA 101, 2003 ed.

Mechanical Advantage (M/A). A force created through mechanical means including, but not limited to, a system of levers, gearing, or ropes and pulleys usually creating an output force greater than the input force and expressed in terms of a ratio of output force to input force. NFPA 1670, 2004 ed.

Mechanical Ventilation. A process of removing heat, smoke, and gases from a fire area by using exhaust fans, blowers, air conditioning systems, or smoke ejectors. (See Figure M.4.) NFPA 402, 2002 ed.

Medical Control. The physician providing direction for patient care activities in the prehospital setting. NFPA 473, 2002 ed.

Medical Director. A physician trained in emergency medicine, designated as a medical director for the local EMS agency. (See Figure M.5.) NFPA 450, 2004 ed.

Medical Evaluation. The analysis of information for the purpose of making a determination of medical certification. *Note: Medical evaluation includes a medical examination.* NFPA 1582, 2003 ed.

Medical Examination. An examination performed or directed by the fire department physician. NFPA 1582, 2003 ed.

Medical Surveillance. The ongoing process of medical evaluation of hazardous materials response team members and public safety personnel who respond to a hazardous materials incident. NFPA 473, 2002 ed.

Medical Transportation Area. That portion of the triage area where injured persons are staged for transportation to medical facilities under the direct supervision of a medical transportation officer. NFPA 424, 2002 ed.

Medical Waste. Items to be disposed of that have been contaminated with human waste, blood, or body fluids, or human waste, human tissue, blood, or body fluids for which special handling precautions are necessary. NFPA 1581, 2005 ed.

Medically Certified. A determination by the fire department physician that the candidate or current member meets medical requirements. NFPA 1582, 2003 ed.

Medium-Expansion Foam. Foams with expansion ratios ranging from 20:1 to 200:1. NFPA 1145, 2000 ed.

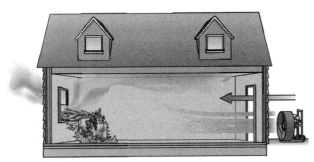

FIGURE M.4 Mechanical Ventilation.

FIGURE M.5 Medical Director.

Melt. A response to heat by a material resulting in evidence of flowing or dripping. NFPA 1971, 2000 ed.

Member. A person involved in performing the duties and responsibilities of a fire department under the auspices of the organization. NFPA 1500, 2002 ed.

Member Assistance Program (MAP). A generic term used to describe the various methods used in the fire department for the control of alcohol and other substance abuse, stress, and personal problems that adversely affect member performance. NFPA 1500, 2002 ed.

Member Organization. An organization formed to represent the collective and individual rights and interests of the members of the fire department, such as a labor union or fire fighters' association. NFPA 1500, 2002 ed.

Mercantile Occupancy. An occupancy used for the display and sale of merchandise. NFPA 5000, 2002 ed.

Metacentric Height. A movable point used to determine stability when related to the center of gravity and center of buoyancy. NFPA 1925, 2004 ed.

Mezzanine. An intermediate level between the floor and the ceiling of any room or space. NFPA 101, 2003 ed.

Microwave. Radio waves with frequencies of 1000 MHz and higher. NFPA 1221, 2002 ed.

Midsagittal Plane. The anatomical plane perpendicular to the basic plane and containing the midpoint of the line connecting the notches of the right and left inferior orbital ridges, and the midpoint of the line connecting the superior rims of the right and left auditory meatus. (See Figure M.6.) NFPA 1971, 2000 ed.

Minimum Continuous Electrical Load. The electrical current required to continuously operate a defined set of electrical devices. NFPA 1901, 2003 ed.

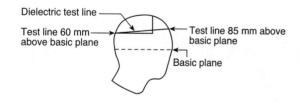

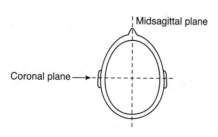

Figure M.6 Midsagittal Plane.

Minimum Ignition Energy (MIE). The minimum energy required from a capacitive spark discharge to ignite the most easily ignitible mixture of a gas or vapor. NFPA 497, 2004 ed.

Minimum Water Supply. The quantity of water required for fire control. NFPA 1142, 2001 ed.

Minor Repair Garage. A building or portions of a building used for lubrication, inspection, and minor automotive maintenance work, such as engine tune-ups, replacement of parts, fluid changes (e.g., oil, antifreeze, transmission fluid, brake fluid, air conditioning refrigerants, etc.), brake system repairs, tire rotation, and similar routine maintenance work, including associated floor space used for offices, parking, or showrooms. NFPA 30A, 2003 ed.

Minor Seams. Seam assemblies that are not classified as Major A or Major B seams. NFPA 1851, 2001 ed.

Miscellaneous Equipment. Portable tools and equipment carried on a fire apparatus not including suction hose, fire hose, ground ladders, fixed power sources, hose reels, cord reels, breathing air systems, or other major equipment or components specified by the purchaser to be permanently mounted on the apparatus as received from the apparatus manufacturer. NFPA 1901, 2003 ed.

Miscellaneous Equipment Allowance. That portion of the GVWR or GCWR allocated for the weight of the miscellaneous equipment and its mounting brackets, boards, or trays. NFPA 1901, 2003 ed.

Miscibility. The property of being capable of mixing in any ratio without separation into phases. NFPA 1150, 2004 ed.

Mitigation. Activities taken to eliminate or reduce the probability of the event, or reduce its severity or consequences, either prior to or following a disaster/emergency. NFPA 1600, 2004 ed.

Mix Ratio. The proportion of foam concentrate in the foam solution, expressed as a volume percentage. NFPA 1150, 2004 ed.

Mixing Chamber. A device used to mix foam solution and air. NFPA 1145, 2000 ed.

Mobile Emergency Hospital (MEH). A specialized, self-contained vehicle that can provide a clinical environment that enables a physician to provide definitive treatment for serious injuries at the accident scene. NFPA 424, 2002 ed.

Mobile Foam Fire Apparatus. Fire apparatus with a permanently mounted fire pump, foam proportioning system, and foam concentrate tank(s) whose primary purpose is for use in the control and extinguishment of flammable and combustible liquid fires in storage tanks and other flammable liquid spills. NFPA 1901, 2003 ed.

Mobile Property Type. Property that was designed to be movable in relation to fixed property regardless of whether the property is currently movable, for example, vehicles, ships, and airplanes. NFPA 901, 2001 ed.

Mobile Unit. A two-way radio-equipped vehicle or person; also a two-way radio by itself that is associated with a vehicle or person. (See Figure M.7.) NFPA 1221, 2002 ed.

Mobile Water Supply Apparatus (Tanker, Tender). A vehicle designed primarily for transporting (pickup, transporting, and delivering) water to fire emergency scenes to be applied by other vehicles or pumping equipment. (See Figure M.8.) NFPA 1901, 2003 ed.

Mode of Transmission. The physical means of entry of a hazardous material into the human body, including inhalation, absorption, ingestion, and injection. (See Figure M.9.) NFPA 1006, 2003 ed.

Model. The collective term used to identify a group of individual elements of the same basic design and components from a single manufacturer produced by the same manufacturing and quality assurance procedures that are covered by the same certification. NFPA 1971, 2000 ed.

Model Weight. The basic weight of the helmet plus accessories for the specific model identified. NFPA 1971, 2000 ed.

Modem (Modulator/Demodulator Unit). A device that converts data that are compatible with data-processing equipment to a form that is compatible with transmission equipment, and vice versa. NFPA 1221, 2002 ed.

Modification. An alteration or adjustment to any component that is a deviation from the original specifications or design of the fire apparatus. NFPA 1915, 2000 ed.

Moisture Barrier. The portion of the composite designed to prevent the transfer of liquids. NFPA 1971, 2000 ed.

FIGURE M.8 Mobile Water Supply Apparatus.

Momentary Switch. A switch that returns to the neutral position (off) when released. NFPA 1901, 2003 ed.

Monitor. *As applied to emergency services communications:* To listen to radio messages without transmitting a response. NFPA 1221, 2002 ed. *As applied to marine vessels:* A fixed master stream device, manually or remotely controlled, or both, capable of discharging large volumes of water or foam. NFPA 1925, 2004 ed.

Monitor Panel. A device that is located at a position remote from the system being monitored (usually at the bridge) and that indicates the condition of the system being monitored. NFPA 1925, 2004 ed.

FIGURE M.7 Mobile Radio.

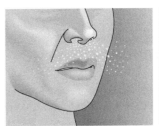 

(a) Inhalation (b) Absorption

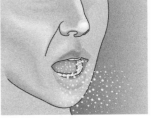 

(c) Ingestion (d) Injection

FIGURE M.9 Mode of Transmission.

Monitoring Equipment. Instruments and devices used to identify and quantify contaminants. NFPA 471, 2002 ed.

Monitoring for Integrity. Automatic monitoring of circuits and other system components for the existence of defects or faults that interfere with receiving or transmitting an alarm. NFPA 1221, 2002 ed.

Moorings. Methods of securing a vessel within a given area. NFPA 1925, 2004 ed.

Mop Up. The act of making a wildland fire scene safer after the fire has been controlled, such as extinguishing or removing burning material along or near the control line, felling snags, and trenching logs to prevent rolling. NFPA 901, 2001 ed.

Motion Detector. A component of PASS that senses movement, or lack of movement, and activates the alarm signal under a specified sequence of events. NFPA 1982, 1998 ed.

Motor Fuel Dispensing Facility. That portion of a property where motor fuels are stored and dispensed from fixed equipment into the fuel tanks of motor vehicles or marine craft or into approved containers, including all equipment used in connection therewith. NFPA 30A, 2003 ed.

Motor Fuel Dispensing Facility Located Inside a Building. That portion of a motor fuel dispensing facility located within the perimeter of a building or building structure that also contains other occupancies. NFPA 30A, 2003 ed.

Motor Vehicle Fluid. A fluid that is a flammable, combustible, or hazardous material, such as crankcase fluids, fuel, brake fluids, transmission fluids, radiator fluids, and gear oil. NFPA 1, 2003 ed.

Motor-Generator. A machine that consists of a generator driven by an electric motor. NFPA 1221, 2002 ed.

Moulage. A reproduction of a skin lesion, tumor, wound, or other pathological state applied for realism to simulate injuries in emergency exercises. NFPA 424, 2002 ed.

Movement Area. That part of an airport to be used for the take-off, landing, and taxiing of aircraft, and the apron(s). NFPA 403, 2003 ed.

MSDS. See Material Safety Data Sheet. NFPA 1006, 2003 ed.

Mucous Membrane. A moist layer of tissue that lines the mouth, eyes, nostrils, vagina, anus, or urethra. NFPA 1581, 2005 ed.

Multiple Casualty. Injury or death of more than one individual in an incident. NFPA 450, 2004 ed.

Multiple Configuration. Variable configurations or positions of the aerial device (e.g., elevation, extension) in which a manufacturer's different rated load capacities are allowed. NFPA 1901, 2003 ed.

Multiple Jacket. A construction consisting of a combination of two separately woven reinforcements (double jacket) or two or more reinforcements interwoven. NFPA 1962, 2003 ed.

Multiple-Row Racks. Racks greater than 3.7 m (12 ft) wide or single- or double-row racks, separated by aisles less than 1.1 m (3.5 ft) wide having an overall width greater than 3.7 m (12 ft). NFPA 13, 2002 ed.

Municipal-Type Water System. A system having water pipes servicing hydrants and designed to furnish, over and above domestic consumption, a minimum of 950 L/min (250 gpm) at 138 kPa (20 psi) residual pressure for a 2-hour duration. NFPA 1141, 2003 ed.

Mutual Aid. Reciprocal assistance by emergency services under a prearranged plan. NFPA 402, 2002 ed.

Mutual Aid Agreement. A pre-arranged agreement developed between two or more entities to render assistance to the parties of the agreement. NFPA 1600, 2004 ed.

Mutual Aid Plan. A plan developed between two or more agencies to render assistance to the parties of the agreement. NFPA 1142, 2001 ed.

Nape Device. A device located below the bitragion inion arc used to aid in helmet retention. NFPA 1971, 2000 ed.

National Association of EMS Physicians (NAEMSP). A national organization of emergency medical physicians and other professionals. NFPA 450, 2004 ed.

National Contingency Plan. Policies and procedures of the federal agency members of the National Oil and Hazardous Materials Response Team. NFPA 471, 2002 ed.

National Defense Area. An area established on nonfederal lands located in the United States, its territories, or its possessions for the purpose of safeguarding classified defense information or protecting Department of Defense (DOD) equipment, material, or both. NFPA 1003, 2005 ed.

National Highway Traffic Safety Administration (NHTSA). The agency under the Department of Transportation that is responsible for preventing motor vehicle injuries. NFPA 450, 2004 ed.

National Hose Thread (NH). A standard screw thread that has dimensions for inside (female) and outside (male) fire hose connections as defined in NFPA 1963, Standard for Fire Hose Connections. NFPA 1901, 2003 ed.

National Institutes of Health (NIH). An agency of the Public Health Service of the Department of Health and Human Services, responsible for promoting the nation's health. NFPA 450, 2004 ed.

National Search and Rescue Plan. A document that identifies responsibilities of U.S. federal agencies and serves as the basis for the National Search and Rescue Manual, which discusses search and rescue organizations, resources, methods, and techniques utilized by the federal government. NFPA 1670, 2004 ed.

National Standard Hose Thread (NH). A standard thread that has dimensions for inside and outside fire hose connection screw threads as defined by NFPA 1963, Standard for Fire Hose Connections. NFPA 1963, 1998 ed.

National Transportation and Safety Board (NTSB). A federal agency that is responsible for investigating and determining the probable cause of aircraft accidents. NFPA 402, 2002 ed.

Natural Barricade. A natural outdoor feature(s), such as hills or trees, with a density sufficient to prevent surrounding exposures that require protection from being seen from a magazine or building containing explosives when the trees are bare of leaves. NFPA 1124, 2003 ed.

Neck Circumference. Upper torso measurement from folded edge to folded edge at the midpoint of the collar width with the garment front closure closed at the top and the top edges of the collar in horizontal alignment, and multiplied by 2 to obtain the circumference. NFPA 1977, 2005 ed.

Needle. A slender, usually sharp, pointed instrument used for puncturing tissues, suturing, drawing blood, or passing a ligature around a vessel. NFPA 1581, 2005 ed.

Negative Pressure. Pressure less than atmospheric. NFPA 99, 2005 ed.

Negative Pressure SCBA. An SCBA in which the pressure inside the facepiece, in relation to the pressure surrounding the outside of the facepiece, is negative during any part of the inhalation or exhalation cycle. NFPA 1981, 2002 ed.

Neglect. Failure to act on behalf of or in protection of an individual in one's care. NFPA 1035, 2005 ed.

Nesting. A method of securing cylinders upright in a tight mass using a contiguous three-point contact system whereby all cylinders in a group have a minimum of three contact points with other cylinders or a solid support structure (for example, a wall or railing). NFPA 55, 2005 ed.

Net Pump Pressure. The sum of the discharge pressure and the suction lift converted to psi or kPa when pumping at draft, or the difference between the discharge pressure and the intake pressure when pumping from a hydrant or other source of water under positive pressure. NFPA 1901, 2003 ed.

Neutral Conductor. The grounded current-carrying conductor of all electrical circuits. NFPA 1901, 2003 ed.

Neutral Position. The position of operating controls when the controls are not engaged. NFPA 1914, 2002 ed.

Neutralization. The process of applying acids or bases to a corrosive product to form a neutral salt. NFPA 471, 2002 ed.

NIOSH. National Institute for Occupational Safety and Health of the U.S. Department of Health and Human Services. NFPA 1404, 2002 ed.

NIOSH Approved. Tested and certified by the National Institute for Occupational Safety and Health (NIOSH) of the U.S. Department of Health and Human Services. NFPA 1404, 2002 ed.

NIOSH Certified. Tested and certified by the National Institute for Occupational Safety and Health (NIOSH) of the U.S. Department of Health and Human Services in accordance with the requirements of 42 CFR 84, Subpart H. NFPA 1981, 2002 ed.

No-Load Condition. The status of an engine with standard accessories operating without an imposed load, with the vehicle drive clutches and any special accessory clutches in a disengaged or neutral condition. NFPA 414, 2001 ed.

Non-Fire Service Personnel. All persons, including police, utility company employees, non-fire service medical personnel, and civilians, who are involved with an incident but who are not fire service personnel. NFPA 901, 2001 ed.

Nonaddressable Public Alerting System (NPAS). A system that transmits alerts to nondesignated recipients or locations. NFPA 1221, 2002 ed.

Nonbulk Packaging. Any packaging having a liquid capacity of 450 L (119 gal) or less, a solids capacity of 400 kg (882 lb) or less, or a compressed gas water capacity of 454 kg (1001 lb) or less. NFPA 472, 2002 ed.

Noncombustible Material. A substance that will not ignite and burn when subjected to a fire. NFPA 220, 1999 ed.

Nondedicated Smoke Control System. A smoke control system that shares components with some other system(s), such as the building HVAC system, which changes its mode of operation to achieve the smoke control objective. NFPA 1, 2003 ed.

Nondestructive Testing (NDT). One of several methods used to inspect a structural component without physically altering or damaging the materials. NFPA 1914, 2002 ed.

Nonencapsulating. A type of ensemble that provides liquid splash protection but does not provide vapor- or gas-tight protection, or liquid-tight protection, and does not cover the wearer's respirator. NFPA 1992, 2005 ed.

Nonflammable. (1) Not readily capable of burning with a flame. (2) Not liable to ignite and burn when exposed to flame. *Note: The antonym of nonflammable is flammable.* NFPA 921, 2004 ed.

Nonflammable Gas. A gas that does not meet the definition of a flammable gas. NFPA 55, 2005 ed.

Nonintersecting Trench. A trench cut in a straight or nearly straight line with no crossing or converging trench legs or cuts. NFPA 1006, 2003 ed.

Nonliquefied Compressed Gases. Gases, other than those in solution, that are contained in a packaging under the charged pressure and are entirely gaseous at a temperature of 20°C (68°F). NFPA 1, 2003 ed.

Nonprimary Protective Garment. A garment that is designed, certified, and intended to be the barrier of protection from a specific hostile environment. NFPA 1975, 2004 ed.

Nonremovable SCBA-Integrated PASS. An SCBA-Integrated PASS that is not designed and not intended to be readily removed from the PASS/SCBA device so that it cannot be used independently of the SCBA. NFPA 1982, 1998 ed.

Nonthreaded Coupling or Adapter. A coupling or adapter in which the mating is achieved with locks or cams but without the use of screw threads. NFPA 1963, 1998 ed.

Normal Temperature and Pressure (NTP). A temperature of 21.1°C (70°F) and a pressure of 1 atmosphere [101.3 kPa (14.7 psia)]. NFPA 5000, 2002 ed.

Normalized Breakthrough Detection Time. The time at which the permeation rate of a chemical through a material reaches $0.1 \text{ g/cm}^2/\text{min}$. NFPA 1991, 2005 ed.

Nose Gear. That mechanical part of a landing gear system mounted under the nose of an aircraft and designed either as a stationary component or one that retracts into the fuselage. NFPA 402, 2002 ed.

Not Occupied. An area with no persons present; contents or equipment present indicates that the structure is not vacant. NFPA 901, 2001 ed.

Notification. The time at which an alarm is received and acknowledged at a communications center. NFPA 1221, 2002 ed.

Nozzle. A constricting appliance attached to the end of a fire hose or monitor to increase the water velocity and form a stream. (See Figure N.1.) NFPA 1965, 2003 ed.

Nozzle Pressure. The normal pressure measured at the inlet of the nozzle. NFPA 1964, 2003 ed.

FIGURE N.1 Nozzle.

Nozzle Reaction. Force that occurs when a water stream is discharged from the nozzle. NFPA 1901, 2003 ed.

NPSH (National Pipe Straight Hose Thread). National pipe straight hose thread as specified in ASME B1.20.7, Hose Coupling Screw Threads, Inch. NFPA 1906, 2001 ed.

Nursing Home. A building or part of a building used on a 24-hour basis for the housing and nursing care of four or more persons who, because of mental or physical incapacity, might be unable to provide for their own needs and safety without the assistance of another person. NFPA 5000, 2002 ed.

NWCG. National Wildland Fire Coordinating Group. NFPA 1051, 2002 ed.

Objective. A goal that is achieved through the attainment of a skill, knowledge, or both, and that can be observed or measured. NFPA 472, 2002 ed.

Occupancy Hazard Classification Number. A series of numbers from 3 through 7 that are mathematical factors used in a formula to determine total water supply requirements. NFPA 1142, 2001 ed.

Occupant Load. The total number of persons that might occupy a building or portion thereof at any one time. NFPA 5000, 2002 ed.

Occupant-Use Hose. Fire hose designed to be used by the building's occupants to fight incipient fires prior to the arrival of trained fire fighters or fire brigade members. NFPA 1961, 2002 ed.

Occupational Exposure. An exposure incident that resulted from performance of a member's duties. NFPA 1581, 2005 ed.

Occupational Illness. An illness or disease contracted through or aggravated by the performance of the duties, responsibilities, and functions of a fire department member. NFPA 1500, 2002 ed.

Occupational Injury. An injury sustained during the performance of the duties, responsibilities, and functions of a fire department member. NFPA 1500, 2002 ed.

Occupational Safety and Health Program. The overall program to provide occupational safety and health in a fire department as defined in NFPA 1500, Standard on Fire Department Occupational Safety and Health Program. NFPA 1521, 2002 ed.

Occupied. An area with persons present. NFPA 901, 2001 ed.

Off. A functional mode in which the PASS is deactivated. NFPA 1982, 1998 ed.

Off-Pavement Performance. A vehicle's ability to perform or operate on other than paved surfaces. NFPA 414, 2001 ed.

Off-Road Use. Use of fire department vehicles in areas where there is a need to traverse steep terrain or to cross natural hazards on or protruding from the ground. NFPA 1002, 2003 ed.

Off-Road Use Vehicle. A vehicle designed to be used on other than paved or improved roads, especially in areas where no roads, poor roads, and steep grades exist and where natural hazards, such as rocks, stumps, and logs, protrude from the ground. NFPA 1906, 2001 ed.

Offensive Fire Fighting. The mode of manual fire control in which manual fire suppression activities are concentrated on reducing the size of a fire to accomplish extinguishment. NFPA 600, 2005 ed.

Offensive Operations. Actions generally performed in the interior of involved structures that involve a direct attack on a fire to directly control and extinguish the fire. NFPA 1500, 2002 ed.

Officer. The member who is assigned by the incident commander or by any other person of comparable responsibility in the emergency service organization's incident management system. (See Figure O.1.) NFPA 1561, 2005 ed.

Ohm. The unit of electrical resistance (R) that measures the resistance between two points of a conductor when a constant difference of potential of one volt between these two points produces in this conductor a current of one ampere. NFPA 921, 2004 ed.

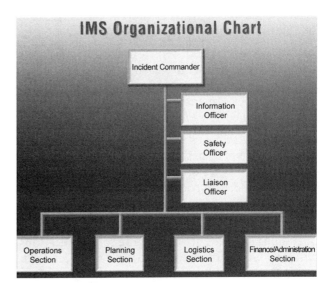

Figure O.1 Incident Management System Organization Chart.

One- and Two-Family Dwelling. Buildings containing not more than two dwelling units in which each dwelling unit is occupied by members of a single family with not more than three outsiders, if any, accommodated in rented rooms. NFPA 1, 2003 ed.

One-Call Utility Location Service. A service from which contractors, emergency service personnel, and others can obtain information on the location of underground utilities in any area. NFPA 1670, 2004 ed.

Open Parking Structure. A parking structure that, at each parking level, has wall openings open to the atmosphere, for an area of not less than 1.4 ft^2 for each linear foot (0.4 m^2 for each linear meter) of its exterior perimeter. Such openings are distributed over 40 percent of the building perimeter or uniformly over two opposing sides. Interior walls lines and column lines are at least 20 percent open, with openings distributed to provide ventilation. NFPA 5000, 2002 ed.

Open System Use. Use of a solid or liquid hazardous material in a vessel or system that is continuously open to the atmosphere during normal operations and where vapors are liberated or the product is exposed to the atmosphere during normal operations. NFPA 1, 2003 ed.

Operating Pressure. *As applied to fire hose:* The highest pressure at which the hose should be used in regular operation. NFPA 1961, 2002 ed. *As applied to fire science:* The pressure at which a system operates. NFPA 1, 2003 ed.

Operating Unit (Vessel) or Process Unit (Vessel). The equipment in which a unit operation or unit process is conducted. NFPA 30, 2003 ed.

Operational Cycle. The movement of the powered rescue tool from the fully closed or retracted position to the fully open or extended position and returned to the fully closed or retracted position. NFPA 1936, 2005 ed.

Operational Tests. An all-vehicle test conducted by the manufacturer to ensure that each vehicle is fully operational when it is delivered and to ensure that the original level of performance of the prototype vehicle has been maintained. NFPA 414, 2001 ed.

Operations. *As applied to flammable and combustible liquids:* A general term that includes, but is not limited to, the use, transfer, storage, and processing of liquids. NFPA 30, 2003 ed. *As applied to wildland fire management:* The incident management section responsible for all tactical operations at the incident. NFPA 1143, 2003 ed.

Operations Room. The room in the communications center where alarms are received and retransmitted. NFPA 1221, 2002 ed.

Operator. *As applied to fire department aerial devices:* A person qualified to operate an aerial device. NFPA 1914, 2002 ed. *As applied to fire apparatus:* A person qualified to operate a fire apparatus. NFPA 1912, 2001 ed.

Operator Alert Device. Any device, whether visual, audible, or both, installed in the driving compartment or at an operator's panel, to alert the operator to either a pending failure, an occurring failure, or a situation that requires his or her immediate attention. NFPA 1915, 2000 ed.

Operator's Panel. A panel containing gauges, switches, instruments, or controls where an operator can visually monitor the applicable functions. NFPA 1901, 2003 ed.

Optical Center. The point specified by the optical warning device manufacturer of highest intensity when measuring the output of an optical warning device. NFPA 1901, 2003 ed.

Optical Element. Any individual lamp or other light emitter within an optical source. NFPA 1901, 2003 ed.

Optical Power. A unit of measure designated as candela-seconds/minute that combines the flash energy and flash rate of an optical source into one power measurement representing the true visual effectiveness of the emitted light. NFPA 1901, 2003 ed.

Optical Source. Any single, independently mounted, light-emitting component in a lighting system. NFPA 1901, 2003 ed.

Optical Warning Device. A manufactured assembly of one or more optical sources. NFPA 1901, 2003 ed.

Ordinary Hazard. Contents that are likely to burn with moderate rapidity or to give off a considerable volume of smoke. NFPA 520, 2005 ed.

Ordinary (Moderate) Hazard. An occupancy in which the total amount of Class A combustibles and Class B flammables are present in greater amounts than expected under light (low) hazard occupancies. *Note: Ordinary (moderate) hazard occupancies could consist of dining areas, mercantile shops, and allied storage; light manufacturing, research operations, auto showrooms, parking garages, workshop or support service areas of light (low) hazard occupancies; and warehouses containing Class I or Class II commodities as defined by NFPA 13, Standard for the Installation of Sprinkler Systems.* (See Figure O.2.) NFPA 1, 2003 ed.

Organic Peroxide. Any organic compound having a double oxygen or peroxy (-O-O-) group in its chemical structure. NFPA 432, 2002 ed.

FIGURE O.2 Example of Ordinary Hazard Area.

Organic Peroxide Formulation. A pure organic peroxide or a mixture of one or more organic peroxides with one or more other materials in various combinations and concentrations. NFPA 432, 2002 ed.

Organization. The entity that provides the direct management and supervision for the emergency incident response personnel. NFPA 1851, 2001 ed.

Organization's Area of Specialization. Any chemicals and containers used by the private sector specialist employee's employer. NFPA 472, 2002 ed.

OSHA. The Occupational Safety and Health Administration of the U.S. Department of Labor. NFPA 55, 2005 ed.

Other Gas. A gas that is not a corrosive gas, flammable gas, highly toxic gas, oxidizing gas, pyrophoric gas, toxic gas, or unstable reactive gas with a hazard rating of Class 2, Class 3, or Class 4 gas, that might be a nonflammable gas or inert gas. NFPA 1, 2003 ed.

Out of Service. Resources assigned to an incident but unable to respond for mechanical, rest, or personnel reasons. NFPA 1051, 2002 ed.

Outcome. The result, effects, or consequences of an emergency system encounter on the health status of the patient. NFPA 450, 2004 ed.

Outdoor Area. An area that is either outside the confines of a building, or an area sheltered from the elements by overhead cover, that is protected from weather exposure by an exterior wall that obstructs not more than 25 percent of the building boundary. NFPA 1, 2003 ed.

Outer Boot. A secondary boot worn over the footwear ensemble element or bootie for the purpose of providing physical protection. NFPA 1991, 2005 ed.

Outer Garment. *As applied to vapor-protective ensembles:* A secondary garment worn over the suit ensemble element for the purpose of providing physical protection. NFPA 1991, 2005 ed. *As applied to liquid-splash protective ensembles:* A secondary garment worn over another garment that provides physical protection for the chemical-protective material. NFPA 1992, 2005 ed.

Outer Glove. *As applied to vapor-protective ensembles:* A secondary glove worn over the glove ensemble element for the purpose of providing physical protection. NFPA 1991, 2005 ed. *As applied to liquid-splash protective ensembles:* A secondary glove worn over another glove that provides physical protection for the chemical-protective material. NFPA 1992, 2005 ed.

Outer Perimeter. That area outside of the inner perimeter that is secured for immediate-support operational requirements, free of unauthorized or uncontrolled interference. NFPA 424, 2002 ed.

Outer Shell. The outermost layer of the composite with the exception of trim, hardware, reinforcing material, and wristlet material. NFPA 1971, 2000 ed.

Overall Height, Length, and Width. The dimensions determined with the vehicle fully loaded and equipped, unless otherwise specified. NFPA 414, 2001 ed.

Overcurrent. Any current in excess of the rated current of equipment or the ampacity of a conductor. It may result from overload, short circuit, or ground fault. NFPA 70, 2005 ed.

Overfilling Prevention Device. A safety device that is designed to provide an automatic means to prevent the filling of a container in excess of the maximum permitted filling limit. NFPA 58, 2004 ed.

Overhaul. *As applied to aircraft rescue:* A firefighting term involving the process of final extinguishment after the main body of a fire has been knocked down; all traces of fire must be extinguished at this time. NFPA 402, 2002 ed. *As applied to fire apparatus:* To inspect, identify deficiencies, and make necessary repairs to return a component to operational condition. NFPA 1915, 2000 ed.

Overhead. A vessel equivalent to a ceiling. NFPA 1405, 2001 ed.

Overheat. Destruction of material by heat without self-sustained combustion. NFPA 901, 2001 ed.

Override. A system or device used to neutralize a given action or motion. NFPA 1901, 2003 ed.

Override (Aerial Device). The takeover of all aerial device movement control functions by an operator at a second control station. NFPA 1901, 2003 ed.

Oxidizer. Any material that readily yields oxygen or other oxidizing gas, or that readily reacts to promote or initiate combustion of combustible materials and can undergo a vigorous self-sustained decomposition due to contamination or heat exposure. NFPA 430, 2002 ed.

Oxidizing Gas. A gas that can support and accelerate combustion of other materials. NFPA 55, 2005 ed.

Oxygen. A chemical element that, at normal atmospheric temperatures and pressures, exists as a colorless, odorless, and tasteless gas and comprises about 21 percent by volume of the earth's atmosphere. NFPA 53, 2004 ed.

Oxygen Deficiency. Insufficiency of oxygen to support combustion. NFPA 921, 2004 ed.

Oxygen-Deficient Atmosphere. Air atmospheres containing less than 19.5 percent oxygen by volume at one standard atmosphere pressure. NFPA 1500, 2002 ed.

Oxygen-Enriched Atmosphere (OEA). Air atmospheres containing more than 23.5 percent oxygen by volume at one standard atmosphere pressure. NFPA 1670, 2004 ed.

Package. The wrapping or enclosure directly containing a glove or face protection device. NFPA 1999, 2003 ed.

Package Product Label. The product label that is printed on or attached to a package containing one or more compliant products. NFPA 1999, 2003 ed.

Packaging. *As applied to hazardous materials incidents:* Any container that holds a material (hazardous and nonhazardous). NFPA 472, 2002 ed. *As applied to rescue technician qualifications:* The process of securing a victim in a transfer device, with regard to existing and potential injuries or illness, so as to prevent further harm during movement. NFPA 1006, 2003 ed.

Pager. A compact radio receiver used for providing one-way communication or limited digital/data two-way communication. NFPA 1221, 2002 ed.

Panel Team. The group of individuals, with established communications and leadership, assigned to construct (if necessary), move, place, and manage panels (traditional sheeting panels) both inside and outside the space, trench, or excavation. NFPA 1670, 2004 ed.

Paramedic. A medical technician who has received extensive training in advanced life support and emergency medicine. NFPA 424, 2002 ed. *Note: These personnel are usually permitted to administer intravenous fluids and other drugs that can arrest a life-threatening physiological condition.* NFPA 424, 2002 ed.

Parbuckling. A technique for moving a load utilizing a simple 2:1 mechanical advantage system in which the load is placed inside a bight formed in a length of rope, webbing, tarpaulin, blanket, netting, and so forth that creates the mechanical advantage, rather than being attached to the outside of the bight with ancillary rope rescue hardware. NFPA 1006, 2003 ed.

Parenteral. Piercing of the mucous membranes or the skin barrier due to such events as needle sticks, human bites, cuts, and abrasions. NFPA 1581, 2005 ed.

Parking Structure. A building, structure, or portion thereof used for the parking, or storage, or both, of motor vehicles. NFPA 1, 2003 ed.

Participant. Any student, instructor, safety officer, visitor, or other person who is involved in the live fire training evolution within the operations area. NFPA 1403, 2002 ed.

Particulates. Solid matter that is dispersed in air as a mixture. NFPA 1994, 2001 ed.

PASS. See Personal Alert Safety System. NFPA 1982, 1998 ed.

PASS/SCBA Device. The entire assembled equipment that integrally combines a PASS into an SCBA, where the SCBA-Integrated PASS is removable or nonremovable. (See Figure P.1.) NFPA 1982, 1998 ed.

Passageway. A corridor or hallway. NFPA 1405, 2001 ed.

Passive Search Measures. Search efforts that do not require active searching by the rescuers. NFPA 1006, 2003 ed.

Patch Kettle. Any pot or container with a capacity of less than 6 gal (22.7 L) used for preheating tar, asphalt, pitch, or similar substances for the repair of roofs, streets, floors, pipes, or similar objects. NFPA 1, 2003 ed.

Patch Pocket(s). Pockets located on the exterior of protective garments. NFPA 1977, 2004 ed.

Pathogens. Microorganisms such as a bacteria, virus, or fungus that is capable of causing disease. NFPA 1581, 2005 ed.

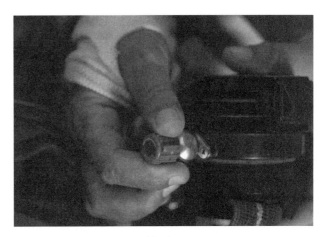

Figure P.1 PASS/SCBA Integrated Device.

Patient. An individual, living or dead, whose body fluids, tissues, or organs could be a source of exposure to the member. NFPA 1581, 2005 ed.

Patient Access Interval. A measurement that begins when the unit comes to a complete stop at the location of the incident and ends when personnel make contact with the patient. NFPA 450, 2004 ed.

Patrol. To systematically observe and check a length of control line during or after its construction to prevent breakovers (slopovers), control spot fires, or extinguish overlooked hot spots. NFPA 1051, 2002 ed.

Pawls. Devices attached to a fly section(s) to engage ladder rungs near the beams of the section below for the purpose of anchoring the fly section(s). NFPA 1931, 2004 ed.

Penetrating Nozzle. An appliance designed to penetrate the skin of an aircraft and inject extinguishing agent. NFPA 402, 2002 ed.

Penetration. The movement of a material through a suit's closures, such as zippers, buttonholes, seams, flaps, or other design features of chemical-protective clothing, and through punctures, cuts, and tears. (See Figure P.2.) NFPA 471, 2002 ed.

Percent Grade. The ratio of the change in elevation to the horizontal distance traveled multiplied by 100. NFPA 414, 2001 ed.

Percent Inward Leakage. The ratio of vapor concentration inside the ensemble versus the vapor concentration outside the ensemble expressed as a percentage. NFPA 1994, 2001 ed.

Performance. Those criteria that are required by members to safely and efficiently do the required essential job tasks. NFPA 1582, 2003 ed.

Peril. An active cause of loss, such as a hurricane, fire, or accident. NFPA 1250, 2004 ed.

Periodic. Occurring or recurring at regular intervals, as determined by the individual organization (e.g., weekly, monthly, quarterly, semiannually, yearly). NFPA 1401, 2001 ed.

Permanent Deformation. That deformation remaining in any part of a ladder or its components after all test loads have been removed from the ladder. NFPA 1931, 2004 ed.

Permeation. A chemical action involving the movement of chemicals, on a molecular level, through intact material. (See Figure P.3.) NFPA 471, 2002 ed.

Permissible Exposure Limit (PEL). The maximum permitted 8-hour, time-weighted average concentration of an airborne contaminant. NFPA 5000, 2002 ed.

Permit. A document issued by the authority having jurisdiction for the purpose of authorizing performance of a specified activity. NFPA 1, 2003 ed.

Peroxide-Forming Chemical. A chemical that, when exposed to air, forms explosive peroxides that are shock sensitive, pressure sensitive, or heat sensitive. NFPA 1, 2003 ed.

Person. An individual, a firm, a copartnership, a corporation, a company, an association, or a joint-stock association, including any trustee, receiver, assignee, or personal representative thereof. NFPA 5000, 2002 ed.

Personal Alert Safety System (PASS). A device that senses movement or lack of movement, and that automatically activates an audible alarm signal (which can also be manually activated) to alert and to assist others in locating a fire fighter or emergency services person who is in danger. (See Figure P.4.) NFPA 1982, 1998 ed.

Personal Care. The care of residents who do not require chronic or convalescent medical or nursing care. NFPA 101, 2003 ed.

Personal Flotation Device (PFD). A displacement device worn to keep the wearer afloat in water. NFPA 1925, 2004 ed.

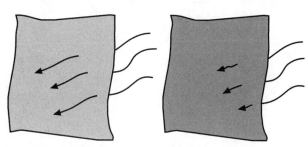

FIGURE P.2 Penetration.

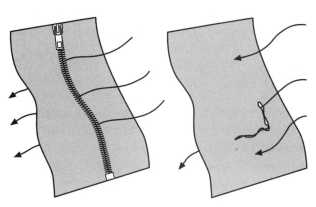

FIGURE P.3 Permeation.

FIGURE P.4 Personal Alert Safety System.

Personal Protective Clothing. *As applied to fire inspectors:* Clothing provided for the fire inspector's personal protection, including a helmet/hard hat, safety glasses, safety shoes/boots, gloves, and coveralls. NFPA 1031, 2003 ed. *As applied to fire fighters:* The full complement of garments fire fighters are normally required to wear while on emergency scene including turnout coat, protective trousers, fire fighting boots, fire fighting gloves, a protective hood, and a helmet with eye protection. (See Figure P.5.) NFPA 1001, 2002 ed.

Personal Protective Equipment. *As applied to wildland fire fighting:* The basic protective equipment for wildland fire suppression, including a helmet, protective footwear, gloves, and flame-resistant clothing as defined in NFPA 1977, Standard on Protective Clothing and Equipment for Wildland Fire Fighting. NFPA 1051, 2002 ed. *As applied to responders to hazardous materials incidents:* The equipment provided to shield or isolate a person from the chemical, physical, and thermal hazards that can be encountered at a hazardous materials incident. NFPA 472, 2002 ed. *As applied to infection control:* Specialized clothing or equipment worn by a member for protection against a hazard. NFPA 1581, 2005 ed. *As applied to search and rescue incidents:* The equipment provided to shield or isolate personnel from infectious, chemical, physical, and thermal hazards. NFPA 1670, 2004 ed.

Personnel. Any individual participating within the incident scene. NFPA 1670, 2004 ed.

Personnel Accountability System. A system that readily identifies both the location and function of all members operating at an incident scene. (See Figure P.6.) NFPA 1500, 2002 ed.

FIGURE P.5 Personal Protective Clothing.

Physical Hazard. A chemical for which there is scientifically valid evidence that it is an organic peroxide or oxidizer. NFPA 1, 2003 ed.

Physical Hazard Material. A chemical or substance classified as a combustible liquid, combustible fiber, explosive, flammable cryogen, flammable gas, flammable liquid, flammable solid, organic peroxide, oxidizer, oxidizing cryogen, pyrophoric, unstable (reactive), or water-reactive material. NFPA 5000, 2002 ed.

Pink Noise. Noise that contains constant energy per octave band. NFPA 1981, 2002 ed.

Pinrail. A rail on or above a stage through which belaying pins are inserted and to which lines are fastened. NFPA 101, 2003 ed.

Pitch Pocket Wood Irregularity. An opening extending parallel to the annual growth rings that contains, or that has contained, either solid or liquid pitch. NFPA 1931, 2004 ed.

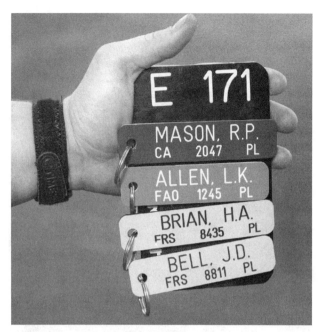

FIGURE P.6 Personnel Accountability Tags.

Plan. A graphic representation of a building structure or portion of a building structure, fire protection system, or fire assembly or equipment. NFPA 1031, 2003 ed.

Plan Developer. The individual, group, or agency responsible for developing and maintaining the preincident plan. NFPA 1620, 2003 ed.

Plan Examiner I. An individual at the first level of progression who has met the job performance requirements specified in this standard for Level I. *Note: The Plan Examiner I conducts basic plan reviews and applies codes and standards.* NFPA 1031, 2003 ed.

Plan Examiner II. An individual at the second or most advanced level of progression who has met the job performance requirements specified in this standard for Level II. *Note: The Plan Examiner II conducts plan reviews and interprets applicable codes and standards.* NFPA 1031, 2003 ed.

Planned Building Groups. Multiple structures constructed on a parcel of land, excluding farmland, under the ownership, control, or development by an individual, a corporation, a partnership, or a firm. NFPA 1141, 2003 ed.

Planned Response. The plan of action, with safety considerations, consistent with the local emergency response plan and an organization's standard operating procedures for a specific hazardous materials incident. NFPA 472, 2002 ed.

Planning. The incident management section responsible for the collection, evaluation, and dissemination of tactical information related to the incident and for preparation and documentation of incident management plans. (See Figure P.7.) NFPA 1143, 2003 ed.

Plant. A facility engaged in the generation and compression of acetylene and in the filling of acetylene cylinders either as its sole operation or in conjunction with facilities for filling other compressed gas cylinders. NFPA 51A, 2001 ed.

Platform. *As applied to marine vessels:* (1) Any flat-topped vessel, such as a barge, capable of providing a working area for personnel or vehicles. (2) A partial deck in the machinery space. NFPA 1405, 2001 ed. *As applied to fire department aerial devices:* An assembly consisting of the support structure, floor, railings, and operator's secondary controls that is attached to the tip of a boom or an aerial ladder for carrying personnel and equipment. (See Figure P.8.) NFPA 1914, 2002 ed.

Plume. The column of hot gases, flames, and smoke rising above a fire; also called convection column, thermal updraft, or thermal column. (See Figure P.9.) NFPA 921, 2004 ed.

Pneumatic Strut. Pneumatic or gas-filled tube and piston assemblies in vehicles or machinery. NFPA 1006, 2003 ed.

Pocket Mask. A double-lumen device that is portable, pocket-size, and designed to protect the emergency care provider from direct contact with the mouth/lips or body fluids of a patient while performing artificial respiration. (See Figure P.10.) NFPA 1581, 2005 ed.

FIGURE P.7 Preincident Plans.

Figure P.8 Platform.

Point of Origin. The exact physical location where a heat source and a fuel come in contact with each other and a fire begins. NFPA 921, 2004 ed.

Policy. A legal agreement for transferring risk that defines what will be paid for, in the event of a defined loss, in exchange for a defined amount of money (premium). NFPA 1250, 2004 ed.

Pompier Ladder (Scaling Ladder). A ladder having a single center beam only, with rungs protruding on either side of the beam, and with a large hook on top that is used for scaling. (See Figure P.11.) NFPA 1931, 2004 ed.

Pool. To join with others in sharing insurance/financial plans and risks. NFPA 1250, 2004 ed.

Port Side. The left-hand side of a ship when facing forward. NFPA 1405, 2001 ed.

Portable Anchor. A manufactured device designed to support human loads. NFPA 1983, 2001 ed.

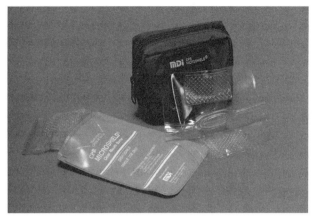

Figure P.10 Pocket Mask.

Portable Electrical Equipment. Any electrical equipment that is not fixed. NFPA 1901, 2003 ed.

Portable Generator. A mechanically driven power source that can be removed from the fire apparatus and operated at a location that is remote from the fire apparatus. NFPA 1901, 2003 ed.

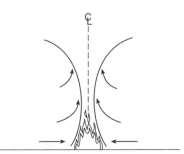

Figure P.9 Plume.

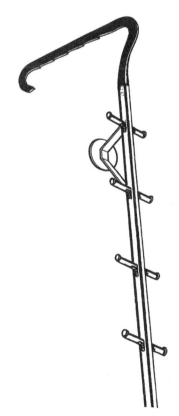

Figure P.11 Pompier Ladder.

Portable Monitor. A monitor that can be lifted from a vehicle-mounted bracket and moved to an operating position on the ground by not more than two people. (See Figure P.12.) NFPA 1965, 2003 ed.

Portable Radio. A battery-operated, hand-held transceiver. (See Figure P.13.) NFPA 1221, 2002 ed.

Portable Tank. Any closed vessel having a liquid capacity over 60 U.S. gal (227 L) and not intended for fixed installation. *Note: This includes intermediate bulk containers (IBCs) as defined and regulated by the U.S. Department of Transportation.* NFPA 1, 2003 ed.

Portable Valve. A fire hose appliance that includes at least one valve and has fire hose connections on both inlet(s) and outlet(s). NFPA 1965, 2003 ed.

Positive Pressure SCBA. An SCBA in which the pressure inside the facepiece, in relation to the pressure surrounding the outside of the facepiece, is positive during both inhalation and exhalation. NFPA 1981, 2002 ed.

Post Aircraft Accident. The specific time when all fires have been extinguished, persons have been accounted for, survivors have been removed, and the hazards have been identified. NFPA 402, 2002 ed.

Post-Exposure Prophylaxis. Administration of a medication to prevent development of an infectious disease following known or suspected exposure to that disease. NFPA 1581, 2005 ed.

Postbriefing. At the termination of an incident, after breakdown and cleanup have occurred, reviews the effectiveness of strategies, tactics, equipment, and personnel at an incident, as well as provides an opportunity to detect the presence of critical incident stress syndrome. NFPA 1006, 2003 ed.

Figure P.13 Portable radio.

Potentially Infectious Materials. Any body fluid that is visibly contaminated with blood; all body fluids in situations where it is difficult or impossible to differentiate between body fluids; sputum, saliva, and other respiratory secretions; and any unfixed tissue or organ from a living or dead human. NFPA 1581, 2005 ed.

Power Source. The power obtained from the utility distribution system, an engine-driven generator, or a battery. NFPA 1221, 2002 ed.

Power Supply Assembly. Any cord or distribution assembly that is partly comprised of the neutral conductor, grounding conductor, and line voltage conductors connected from the output terminals of the power source to the first main overcurrent protection device. NFPA 1901, 2003 ed.

Power Train. The parts of a fire apparatus that transmit power from the engine to the wheels, including the transmission, split shaft power takeoff, midship pump transmission, drive shaft(s), clutch, differential(s), and axles. NFPA 1915, 2000 ed.

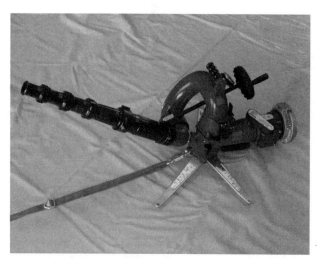

Figure P.12 Portable Ground Monitor.

Power Unit. A powered rescue tool component consisting of a prime mover and the principal power output device used to power the rescue tool. NFPA 1936, 2005 ed.

Power-Assist Steering. A system using hydraulic or air power to aid in the steering assist. *Note: The power-assist steering system is supplementary to the mechanical system in order to maintain steering ability in the event of power failure.* NFPA 414, 2001 ed.

Powered Equipment Rack. A power-operated device that is intended to provide storage of hard suction hoses, ground ladders, or other equipment, generally in a location above apparatus compartments. NFPA 1901, 2003 ed.

Powered Rescue Tool. A rescue tool that receives power from the power unit component and generates the output forces or energy used to perform one or more of the functions of spreading, lifting, holding, crushing, pulling, or cutting. (See Figure P.14.) NFPA 1936, 2005 ed.

Powered Rescue Tool Components. Cable assemblies, hose assemblies, power units, hose reels, and remote valve blocks. *Note: The individual parts that are assembled in a rescue tool or component, such as seals, screws, valves, and switches, are not themselves considered as components for the purposes of this standard.* NFPA 1936, 2005 ed.

ppm. Parts per million, volume per volume. NFPA 1989, 2003 ed.

Practical Critical Fire Area (PCA). This area is two-thirds of the Theoretical Critical Fire Area (TCA). NFPA 402, 2002 ed.

Pre-Alarm Signal. An audible warning that is identifiable as an indication that a PASS is about to sound the alarm signal. NFPA 1982, 1998 ed.

Pre-Entry Briefing. Information passed to all personnel prior to entry into a confined space or trench/excavation environment. NFPA 1670, 2004 ed.

Pre-Entry Medical Exam. A baseline medical evaluation of the rescue entrants performed immediately prior to a rescue entry. NFPA 1006, 2003 ed.

Pre-Incident Plan. A document developed by gathering general and detailed data used by responding personnel to determine the resources and actions necessary to mitigate anticipated emergencies at a specific facility. (See Figure P.15.) NFPA 1620, 2003 ed.

Prebriefing. At the beginning of an incident, after size-up information has been assessed, given to the rescue team to provide assignments, select and notify of strategy and tactics to be performed, and state the mission objective. NFPA 1006, 2003 ed.

Preconnected Hose Line. A hose line that is stored on the apparatus already connected to an outlet on a pump and that can be charged by the activation of one discharge valve. NFPA 1901, 2003 ed.

Figure P.14 Hydraulic Spreader.

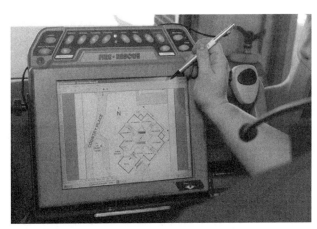

Figure P.15 Pre-Incident Plan.

Preconnected Line. A discharge hose line already attached to an engine outlet. NFPA 1410, 2005 ed.

Premixed Flame. A flame for which the fuel and oxidizer are mixed prior to combustion, as in a laboratory Bunsen burner or a gas cooking range; propagation of the flame is governed by the interaction between flow rate, transport processes, and chemical reaction. NFPA 921, 2004 ed.

Prepackaged Fireworks Merchandise. A consumer fireworks device or group of consumer fireworks devices that has been packaged within an unperforated container or packaging material by the manufacturer, distributor, or seller for retail display and sale as a unit. NFPA 1124, 2003 ed.

Preparation Phase. All actions and planning conducted prior to the initial receipt of alarm. NFPA 1670, 1999 ed.

Prepared Program. An assembled kit, including a lesson plan, behavioral objectives, presentation outline, instructional materials, and evaluation instruments, that is ready to be presented. NFPA 1035, 2005 ed.

Preparedness. Activities, programs, and systems developed and implemented prior to a disaster/emergency that are used to support and enhance mitigation of, response to, and recovery from disasters/emergencies. NFPA 1600, 2004 ed.

Prescribed Fire. Controlled application of fire to wildland fuels in either their natural or modified state, under specified environmental conditions that allow the fire to be confined to a predetermined area and at the same time to produce the intensity of heat and rate of spread required to attain planned resource management objectives. NFPA 901, 2001 ed.

Prescribed Fire (Burning). Any fire ignited by management actions to meet specific objectives. NFPA 1051, 2002 ed.

Preservation of Evidence. After an aircraft accident/incident, it is imperative that investigative evidence be preserved after life safety and rescue operations have been concluded. NFPA 402, 2002 ed.

Pressure. Unless otherwise stated, is expressed in pounds per square inch above atmospheric pressure. NFPA 54, 2002 ed.

Pressurized Aircraft. Sealed, modern-type aircraft within which the internal atmospheric pressure can be regulated. NFPA 402, 2002 ed.

Prevention. Activities, including public education, law enforcement, personal contact, and reduction of fuel hazards, directed at reducing the incidence of fires. NFPA 1143, 2003 ed.

Preventive Maintenance. The act or work of keeping something in proper condition by performing necessary preventive actions, in a routine manner, to prevent failure or breakdown. NFPA 1915, 2000 ed.

Primarily Assigned. The principal fire fighting responsibility in a given jurisdiction, district, or area. NFPA 1500, 2002 ed.

Primary Access. The existing opening of doors and/or windows that provide a pathway to the trapped and/or injured victim(s). NFPA 1670, 2004 ed.

Primary Containment. The first level of containment, consisting of the inside portion of the container that comes into immediate contact on its inner surface with the material being contained. NFPA 1, 2003 ed.

Primary Extinguishing Agent. Agents that have the capability of suppressing and preventing the reignition of fires in liquid hydrocarbon fuels. NFPA 402, 2002 ed.

Primary Inlet. The inlet where an appliance connects to a hose. NFPA 1963, 1998 ed.

Primary Materials. Vapor-protective ensemble and element materials limited to the suit material, hood and visor material, glove material, and footwear material that provide protection from chemical and physical hazards. NFPA 1991, 2005 ed.

Primary Protective Garment. A garment that is designed, certified, and intended to be the barrier of protection from a specific hostile environment. NFPA 1975, 2004 ed.

Primary Suit Materials. Liquid splash-protective ensemble and clothing materials limited to the garment material, hood material, visor material, glove material, and footwear material that provide protection from chemical and physical hazards. This includes, in addition to the above materials, the wearer's respiratory protective equipment when designed to be worn outside the liquid splash-protective ensemble, the umbilical air hose, and all other exposed respiratory equipment materials designed to protect the wearer's breathing air and air path. *Note: Primary materials can be either single layers or composites.* NFPA 1992, 2005 ed.

Primary Turret. The largest capacity foam turret used to apply primary extinguishing agent. NFPA 414, 2001 ed.

Prime Mover. Part of the power unit component; the energy source that drives the principal power output device of the power unit. NFPA 1936, 2005 ed.

Private Sector Specialist Employee A. That person who is specifically trained to handle incidents involving chemicals or containers for chemicals used in the organization's area of specialization. *Note: Consistent with the organization's emergency response plan and standard operating procedures, the private sector specialist employee A is able to analyze an incident involving chemicals within the organization's area of specialization, plan a response to that incident, implement the planned response within the capabilities of the resources available, and evaluate the progress of the planned response.* NFPA 472, 2002 ed.

Private Sector Specialist Employee B. That person who, in the course of regular job duties, works with or is trained in the hazards of specific chemicals or containers within the individual's area of specialization. *Note: Because of the employee's education, training, or work experience, the private sector specialist employee B can be called upon to respond to incidents involving these chemicals or containers. The private sector specialist employee B can be used to gather and record information, provide technical advice, and provide technical assistance (including work within the hot zone) at the incident consistent with the organization's emergency response plan and standard operating procedures and the local emergency response plan.* NFPA 472, 2002 ed.

Private Sector Specialist Employee C. That person who responds to emergencies involving chemicals and/or containers within the organization's area of specialization. *Note: Consistent with the organization's emergency response plan and standard operating procedures, the private sector specialist employee C can be called upon to gather and record information, provide technical advice, and/or arrange for technical assistance. A private sector specialist employee C does not enter the hot or warm zone at an emergency.* NFPA 472, 2002 ed.

Private Street. Any accessway normally intended for vehicular use not dedicated as a public street. NFPA 1141, 2003 ed.

Probability. The likelihood or relative frequency of an event as expressed as a number between 0 and 1. NFPA 1250, 2004 ed.

Procedure. *As applied to emergency services incident management:* An organizational directive issued by the authority having jurisdiction or by the department that establishes a specific policy that must be followed. NFPA 1561, 2005 ed. *As applied to fire brigade qualifications:* The series of actions, conducted in an approved manner and sequence, designed to achieve an intended outcome. NFPA 1081, 2001 ed.

Process. The manufacturing, handling, blending, conversion, purification, recovery, separation, synthesis, or use, or any combination, of a commodity or material. NFPA 1, 2000 ed.

Process and Operations. Include the manufacture, storage, and transportation of goods and chemicals; the storage and dispensing of flammable and combustible liquids, solids, and gases; and the manufacture, use, storage, and transportation of explosives, spray painting, milling, and the like. NFPA 1031, 2003 ed.

Process Hazard Analysis. An analysis of a process or system used to identify potential cause and effect relationships and resultant hazards or system failures. NFPA 1620, 2003 ed.

Product. The compliant proximity protective ensemble or the compliant elements of the proximity protective ensemble. NFPA 1971, 2000 ed.

Product Conformance Verification. A system whereby a product conformance verification organization determines that a manufacturer has demonstrated the ability to produce a product that complies with the requirements of a standard, authorizes the manufacturer to use a label on listed products that comply with the requirements of the standard, and establishes a follow-up program conducted by the product conformance verification organization as a check on the methods the manufacturer uses to determine continued compliance of labeled and listed products with the requirements of the standard. NFPA 1936, 2005 ed.

Product Conformance Verification Organization. An independent third-party organization that determines product compliance with the requirements of a standard with a labeling/listing/follow-up program. NFPA 1936, 2005 ed.

Product Label. A label or marking affixed to a product by the manufacturer that provides general information, warnings, instructions for care and maintenance, and other information. NFPA 1971, 2000 ed.

Production Lot. An identifiable series of products manufactured with identical design specifications and identical materials, and produced without any alterations to technique or procedure. NFPA 1983, 2001 ed.

Professional Architect. An individual technically and legally qualified to practice the profession of architecture. NFPA 1, 2000 ed.

Program. A comprehensive strategy that addresses safety issues via educational means. NFPA 1035, 2005 ed.

Promotion. The advancement of a member from one rank to a higher rank by a method such as election, appointment, merit, or examination. NFPA 1021, 2003 ed.

Proof Load. The application of force to a material as a nondestructive test to verify the performance of that material. NFPA 1983, 2001 ed.

Proof Test Pressure. A pressure equal to at least two times the service test pressure. NFPA 1961, 2002 ed.

Proper(ly). In accordance with the manufacturer's specifications or as recommended by the manufacturer. NFPA 1914, 2002 ed.

Properly. As recommended by the manufacturer. NFPA 1915, 2000 ed.

Property Inventory. Information known about a property before an emergency occurs. NFPA 901, 2001 ed.

Property Use. The use to which a property is put. NFPA 901, 2001 ed.

Proprietary Information. Information regarding compounds or ingredients used in a process or production that do not qualify as trade secrets but that provide an industry or business with a competitive advantage. NFPA 1, 2003 ed.

Protection in Place. The strategy and tactics used to protect or shelter people and/or animals from an advancing wildland fire in a safe area, as an alternative to evacuation. NFPA 1051, 2002 ed.

Protective Clothing Ensemble. Multiple elements of clothing and equipment designed to provide a degree of protection for fire fighters from adverse exposures to the inherent risks of structural fire fighting operations and certain other emergency operations. The elements of the protective ensemble are coats, trousers, coveralls, helmets, gloves, footwear, and interface components. (See Figure P.16.) NFPA 1851, 2001 ed.

Protective Clothing Material. Any material or composite used in an ensemble or ensemble element for the purpose of protecting parts of the wearer's body against chemical/biological terrorism agents, or against physical hazards. NFPA 1994, 2001 ed.

Protective Housing. An enclosure that surrounds the laser or laser system that prevents access to laser radiation above the applicable maximum permissible exposure (MPE) level. NFPA 115, 2003 ed.

Protective System. A method of protecting employees from cave-ins, from material that could fall or roll from an excavation face or into an excavation, or from the collapse of adjacent structures. NFPA 1670, 2004 ed.

Figure P.16 Protective Clothing Ensemble.

Protein Foam (P). A protein-based foam concentrate that is stabilized with metal salts to make a fire-resistant foam blanket. NFPA 403, 2003 ed.

Protocol. A series of sequential steps describing the precise patient treatment. NFPA 473, 2002 ed.

Prototype Vehicle. The first vehicle of a unique vehicle configuration built to establish its performance capability and the performance capability of all subsequent vehicles manufactured from its drawings and parts list. NFPA 414, 2001 ed.

Proximate Cause. The cause that directly produces the effect without the intervention of any other cause. NFPA 921, 2004 ed.

Proximity Fire Fighting. Specialized fire fighting operations that can include the activities of rescue, fire suppression, and property conservation at incidents involving fires producing very high levels of radiant heat as well as conductive and convective heat. NFPA 1977, 2005 ed.

Proximity Personal Protective Equipment (PrPPE). A protective ensemble consisting of approved proximity protective clothing, self-contained breathing apparatus (SCBA), and a personal alert safety system (PASS). NFPA 1003, 2005 ed.

Proximity Protective Coat. A proximity protective garment; an element of the proximity protective ensemble designed to provide minimum protection to upper torso and arms, excluding the hands and head. (See Figure P.17.) NFPA 1971, 2000 ed.

Proximity Protective Coverall. A proximity protective garment; an element of the protective ensemble configured as a single-piece garment and designed to provide minimum protection to the torso, arms, and legs, excluding the head, hands, and feet. NFPA 1971, 2000 ed.

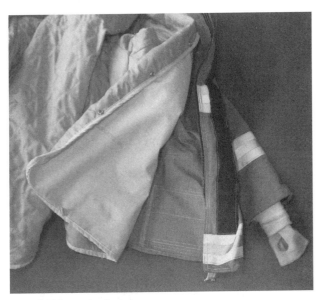

Figure P.17 Protective Coat.

Proximity Protective Ensemble. Multiple elements of clothing and equipment (coats, trousers, coveralls, helmets, gloves, footwear, and interface components) designed to provide a degree of protection for fire fighters from adverse exposures to the inherent risks of proximity fire fighting operations and certain other emergency operations where high levels of radiant heat, as well as convective and conductive heat, are a hazard. NFPA 1971, 2000 ed.

Proximity Protective Footwear. An element of the proximity protective ensemble designed to provide minimum protection to the foot, ankle, and lower leg. NFPA 1971, 2000 ed.

Proximity Protective Garment. The coat, trouser, or coverall elements of the proximity protective ensemble designed to provide minimum protection to the upper and lower torso, arms, and legs, excluding the head, hands, and feet. NFPA 1971, 2000 ed.

Proximity Protective Glove. An element of the proximity protective ensemble designed to provide minimum protection to the fingers, thumb, hand, and wrist. NFPA 1971, 2000 ed.

Proximity Protective Helmet. An element of the proximity protective ensemble designed to provide minimum protection to the head. NFPA 1971, 2000 ed.

Proximity Protective Hood. The interface component element of the proximity protective ensemble designed to provide limited protection to the coat-helmet-SCBA facepiece interface area. NFPA 1971, 2000 ed.

Proximity Protective Trouser. A proximity protective garment; an element of the proximity protective ensemble that is designed to provide minimum protection to the lower torso and legs, excluding the ankles and feet (See Figure P.18.) NFPA 1971, 2000 ed.

psi. Pounds per square inch. NFPA 58, 2004 ed.

psig. Pounds per square inch gauge. NFPA 51, 2002 ed.

PTO. Power takeoff. NFPA 1901, 2003 ed.

Public Alert. A signal or message delivered to a person or device indicating the existence of a situation that affects public safety. NFPA 1221, 2002 ed.

Figure P.18 Protective Trouser.

FIGURE P.19 Prevention Materials.

Public Alerting System (PAS). A system that creates, transmits, and receives a public alert message or signal, or both, that is intended to protect the public from loss of life, health, and property. NFPA 1221, 2002 ed.

Public Fire and Life Safety Education. Comprehensive community fire and injury prevention programs designed to eliminate or mitigate situations that endanger lives, health, property, or the environment. (See Figure P.19.) NFPA 1035, 2005 ed.

Public Fire and Life Safety Education Strategy. An organization's comprehensive plan that is designed, through public fire and life safety education programs, campaigns, and initiatives, to eliminate or mitigate risks that endanger lives, health, property, or the environment. NFPA 1035, 2005 ed.

Public Fire and Life Safety Educator I. The individual who has demonstrated the ability to coordinate and deliver existing educational programs and information. NFPA 1035, 2005 ed.

Public Fire and Life Safety Educator II. The individual who has demonstrated the ability to prepare educational programs and information to meet identified needs. NFPA 1035, 2005 ed.

Public Fire and Life Safety Educator III. The individual who has demonstrated the ability to create, administer, and evaluate educational programs and information. NFPA 1035, 2005 ed.

Public Fire Department. An organization providing rescue, fire suppression, emergency medical services, and related activities to the public. NFPA 1720, 2004 ed.

Public Information Officer. The individual who has demonstrated the ability to conduct media interviews and prepare news releases and media advisories. (See Figure P.20.) NFPA 1035, 2005 ed.

Public Reporting System. A system of alarm-initiating devices, receiving equipment, and connecting circuits, other than a public telephone network, used to transmit alarms from street locations to the communications center. NFPA 1221, 2002 ed.

Public Safety Agency. Organizations providing law enforcement, emergency medical, fire, rescue, communications, and related support services. The term "public safety agency" includes any public, governmental, private, industrial, or military organization engaged in one or more of these activities. NFPA 1061, 2002 ed.

Public Safety Answering Point (PSAP). A facility in which 9-1-1 calls are answered, either directly or though rerouting. NFPA 1221, 2002 ed.

Public Safety Diver. An individual who performs public safety diving. NFPA 1670, 2004 ed.

Public Safety Diving. Underwater diving, related to team operations and training, performed by any member, group, or agency of a community or a government-recognized public safety diving or water rescue team. NFPA 1670, 2004 ed.

Public Safety Element. A section of a land use plan that describes the hazards to public safety and how they are to be mitigated. NFPA 1051, 2002 ed.

Public Safety Telecommunicator Candidate. The person who has fulfilled the entrance requirements of NFPA 1061 but who has not met the requirements of Public Safety Telecommunicator I. NFPA 1061, 2002 ed.

Public Safety Telecommunicator I. The initial contact in managing requests for services who obtains and prepares the pertinent information for the allocation of

FIGURE P.20 Public Information Officer.

public safety resources. *Note: The Public Safety Telecommunicator I makes independent decisions, conveys information, and provides referrals; works in cooperation with a Public Safety Telecommunicator II.* NFPA 1061, 2002 ed.

Public Safety Telecommunicator II. Prioritizes, initiates, and coordinates the response of public safety agencies; manages the flow of incident-related information to and from field units and/or public safety resources; monitors status of field units and assigns additional resources as requested and/or required. NFPA 1061, 2002 ed.

Public Street. A thoroughfare that has been dedicated for vehicular use by the public. NFPA 1141, 2003 ed.

Public Way. A street, alley, or other similar parcel of land essentially open to the outside air deeded, dedicated, or otherwise permanently appropriated to the public for public use and having a clear width and height of not less than 3050 mm (120 in.). NFPA 101, 2003 ed.

Pulley. A device with a free-turning, grooved metal wheel (sheave) used to reduce rope friction with side plates available for a carabiner to be attached. NFPA 1670, 2004 ed.

Pulling Force. The force to pull that is generated by a powered rescue tool and that is measured or calculated at the standard production pulling attachment points on the tool. NFPA 1936, 2005 ed.

Pump Operator. The fire apparatus driver/operator who has met the requirements for the operation of apparatus equipped with an attack or fire pump. NFPA 1002, 2003 ed.

Pump Operator's Panel. The area on a fire apparatus that contains the gauges, controls, and other instruments used for operating the pump. NFPA 1901, 2003 ed.; NFPA 1906, 2001 ed.; NFPA 1912, 2001 ed.

Pump Operator's Position. The location from which the pump operator operates the pump. NFPA 1906, 2001 ed.

Pumper. Fire apparatus with a permanently mounted fire pump of at least 750 gpm (3000 L/min) capacity, water tank, and hose body whose primary purpose is to combat structural and associated fires. (See Figure P.21.) NFPA 1901, 2003 ed.

Pumping System. A pump, the piping, and associated devices mounted permanently on a piece of fire apparatus for the purpose of delivering a fire stream. NFPA 1002, 2003 ed.

Puncture-Resistant Device. The reinforcement to the bottom of footwear located between the sole with heel and the insole that is designed to provide puncture resistance. NFPA 1971, 2000 ed.

Purchaser. The authority having responsibility for the specification and acceptance of the apparatus. NFPA 1901, 2003 ed.

Figure P.21 Pumper.

Purchasing Authority. The agency that has the sole responsibility and authority for negotiating, placing, and, where necessary, modifying each and every solicitation, purchase order, or other award issued by a governing body. NFPA 1906, 2001 ed.

Purification System. A combination of mechanical, chemical, and physical devices such as separators, filters, adsorbents, and catalysts designed to remove or alter contaminants within the compressed air stream to produce effluent air that is breathable. NFPA 1901, 2003 ed.

Pyrolysis. The destructive distillation of organic compounds in an oxygen-free environment that converts the organic matter into gases, liquids, and char. (See Figure P.22.) NFPA 820, 2003 ed.

Pyrophoric. A chemical that spontaneously ignites in air at or below a temperature of 54.5°C (130°F). NFPA 1, 2003 ed.

Pyrophoric Gas. A gas with an autoignition temperature in air at or below 704°C (1300°F). NFPA 1, 2003 ed.

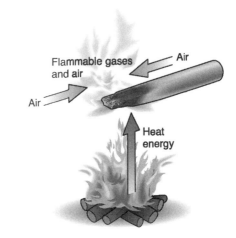

Figure P.22 Pyrolysis.

Qualification. Having satisfactorily completed the requirements of the objectives. NFPA 1021, 2003 ed.

Qualified Person. A person who, by possession of a recognized degree, certificate, professional standing, or skill, and who, by knowledge, training, and experience, has demonstrated the ability to deal with problems relating to a particular subject matter, work, or project. NFPA 1451, 2002 ed.

Quality Assessment (QA). An assessment of the performance of structure, processes, and outcomes within the EMS system and their comparison against a standard. NFPA 450, 2004 ed.

Quality Improvement. The activities undertaken to continuously examine and improve the products and services. NFPA 450, 2004 ed.

Quint Apparatus. A fire department emergency vehicle with a permanently mounted fire pump, a water tank, a hose storage area, an aerial device with a permanently mounted waterway, and a complement of ground ladders. NFPA 1710, 2004 ed.

Rack. Any combination of vertical, horizontal, and diagonal members that supports stored materials. NFPA 230, 2003 ed.

Radiant Heat. Heat energy carried by electromagnetic waves that are longer than light waves and shorter than radio waves; radiant heat (electromagnetic radiation) increases the sensible temperature of any substance capable of absorbing the radiation, especially solid and opaque objects. NFPA 921, 2004 ed.

Radiant Protective Performance (RPP). The resistance of a material to radiant heat, measured in seconds, when exposed to a vertically oriented radiant heat source, positioned at a specific horizontal distance from the vertical placement of the protective material, sufficient to cause a second-degree burn to human tissue. NFPA 1977, 2005 ed.

Radiation. The emission and propagation of energy through matter or space by means of electromagnetic disturbances that display both wave-like and particle-like behavior. (See Figure R.1.) NFPA 801, 2003 ed.

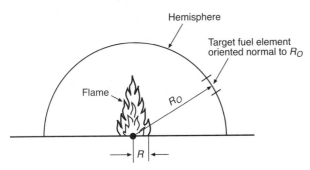

Figure R.1 Radiation.

Radio Channel. A band of frequencies of a width sufficient to allow its use for radio communications. NFPA 72, 2002 ed.

Radio Circuit. A means for carrying out one radio communication at a time, either unidirectionally or bidirectionally. NFPA 1221, 2002 ed.

Radio Communications, Command Channel. A radio channel designated by the emergency services organization that is provided for communications between the incident commander and the tactical level management units during an emergency incident. NFPA 1561, 2005 ed.

Radio Communications, Dispatch Channel. A radio channel designated by the emergency services organization that is provided for communications between the communication center and the incident commander or single resource. NFPA 1561, 2005 ed.

Radio Communications, Tactical Channel. A radio channel designated by the emergency services organization that is provided for communications between resources assigned to an incident, and the incident commander. NFPA 1561, 2005 ed.

Radio Frequency. The number of electromagnetic wave frequency cycles transmitted by a radio in 1 second; specifically, the time taken by a radio signal to complete one cycle. NFPA 1221, 2002 ed.

Radioactive Material. Any material that spontaneously emits ionizing radiation. NFPA 472, 2002 ed.

Radioactive Particulate. Finely divided solids, such as powders and dusts, that emit ionizing radiation in excess of background radiation levels. NFPA 1991, 2005 ed.

Radiography. A nondestructive inspection method that uses x-rays, nuclear radiation, or both to detect discontinuities in material and to present their images on a recording medium. NFPA 1914, 2002 ed.

Radiological Agents. Radiation associated with x-rays, alpha, beta, and gamma emissions from radioactive isotopes or other materials in excess of normal background radiation levels. NFPA 1991, 2000 ed.

Radionuclide. An isotope form of an element or radioactive element that emits radiation in excess of normal background radiation levels. NFPA 1992, 2005 ed.

Raising System. A rope rescue system used to raise a load under control. NFPA 1670, 2004 ed.

Ram. A powered rescue tool that has a piston or other type extender that generates extending forces or both extending and retracting forces. NFPA 1936, 2005 ed.

Ramp. A walking surface that has a slope steeper than 1 in 20. NFPA 101, 2003 ed.

Ramp Breakover Angle. The angle measured between two (2) lines tangent to the front and rear tire static loaded radius, and intersecting at a point on the underside of the vehicle that defines the largest ramp over which the vehicle can drive without the ramp touching the underside of the vehicle. NFPA 1901, 2003 ed.

Rapid Intervention Crew/Company (RIC). A minimum of two fully equipped personnel on site, in a ready state, for immediate rescue of disoriented, injured, lost, or trapped rescue personnel. NFPA 1006, 2003 ed.

Rapid Intervention Crew/Company Universal Air Connection System (RIC UAC). A system that allows emergency replenishment of breathing air to the SCBA of disabled or entrapped fire or emergency services personnel. NFPA 1981, 2002 ed.

Rapid Intervention Team (RIT). Two or more fire fighters assigned outside the hazard area at an interior structure fire to assist or rescue at an emergency operation. NFPA 1410, 2005 ed.

Rate of Perceived Exertion (RPE). A scale created to determine the intensity level of an individual's exertion. *Note: Numeric values are assigned according to the individual's fatigue, environment, muscle factors, etc. It takes into account the subjective aspects of an individual's physical and emotional state, rather than relying solely on an objective percentage of age-predicted maximum heart rate.* NFPA 1584, 2003 ed.

Rated Capacity. The flow rate to which the fire pump manufacturer certifies compliance of the pump when it is new. NFPA 1911, 2002 ed.

Rated Capacity (Aerial Device). The total amount of weight of all personnel and equipment that can be supported at the outermost rung of an aerial ladder or on the platform of an elevating platform with the waterway uncharged. NFPA 1901, 2003 ed.

Rated Capacity (Water Pump). The flow rate to which the pump manufacturer certifies compliance of the pump when it is new. NFPA 1901, 2003 ed.

Rated Discharge. The rate(s) at which a nozzle is designed to flow water when operated at its rated pressure. NFPA 1964, 2003 ed.

Rated Pressure. The pressure at which a nozzle is designed to operate to produce a specified discharge. NFPA 1964, 2002 ed.

Rated Service Time. The period of time, stated on the SCBA's NIOSH certification label, that the SCBA supplied air to the breathing machine when tested to 42 CFR 84, Subpart H. NFPA 1981, 2002 ed.

Rated System Input. The maximum input pressure/electrical power at which the powered rescue tool is designed to operate. NFPA 1936, 2005 ed.

Rated Vertical Height. The vertical distance measured by a plumb line from the maximum elevation of the aerial device allowed by the manufacturer to the ground. NFPA 1914, 2002 ed.

"Reach, Throw, Row, Go." The four sequential steps in water rescue with progressively more risk to the rescuer. (See Figure R.2.) *Note: Specifically, a "go" rescue involves physically entering the medium (e.g., in the water or on the ice).* NFPA 1670, 2004 ed.

Reach/Extension Device. Any device for water rescue that can be extended to a person in the water so that he or she can grasp it and be pulled to safety without physically contacting the rescuer. NFPA 1006, 2003 ed.

Reactive Material. A material that, by itself, is readily capable of detonation, explosive decomposition, or explosive reaction at normal or elevated temperatures and pressures. NFPA 45, 2004 ed.

Readily Accessible. Capable of being reached quickly for operation, renewal, or inspections, without requiring those to whom ready access is requisite to climb over or remove obstacles or to resort to portable ladders, etc. NFPA 70, 1999 ed.

Rebuild. To make extensive repairs in order to restore a component to like-new condition in accordance with the original manufacturer's specifications. NFPA 1071, 2000 ed.

Recall System. The action taken by which a manufacturer identifies an element, provides notice to the users, withdraws an element from the marketplace and distribution sites, and the element is returned to the manufacturer or other acceptable location for corrective action. NFPA 1971, 2000 ed.

Figure R.2 Water Rescue.

Recidivism. Recurrence of firesetting behavior. NFPA 1035, 2005 ed.

Recommended Practice. A document that is similar in content and structure to a code or standard but that contains only nonmandatory provisions using the word "should" to indicate recommendations in the body of the text. NFPA Official Definition.

Record. A permanent account of known or recorded facts that is utilized to recall or relate past events or acts of an organization or the individuals therein. NFPA 1401, 2001 ed.

Recovery. *As applied to rescue:* Nonemergency operations taken by responders to retrieve property or remains of victims. NFPA 1006, 2003 ed. *As applied to disaster/emergency management:* Activities and programs designed to return conditions to a level that is acceptable to the entity. NFPA 1600, 2004 ed.

Recovery Mode. Level of operational urgency where there is no chance of rescuing a victim alive. NFPA 1670, 1999 ed.

Recovery Operations. Those nonemergency activities directed at retrieving property or the remains of victims. NFPA 1951, 2005 ed.

Recreational Fire. The noncommercial burning of materials other than rubbish for pleasure, religious, ceremonial, cooking, or similar purposes in which the fuel burned is not contained in an incinerator, a barbecue grill, or a barbecue pit, and the total fuel area is not exceeding 3 ft (0.9 m) in diameter and 2 ft (0.6 m) in height. NFPA 1, 2003 ed.

Recruit. An individual who has passed beyond the candidate level and who has actively commenced duties as a member of the organization. NFPA 1404, 2002 ed.

Rectifier. A device without moving parts that changes alternating current to direct current. NFPA 1221, 2002 ed.

Reduced Flow Valve. A valve equipped with a restricted flow orifice that is designed to reduce the maximum flow from the valve under full flow conditions. NFPA 1, 2003 ed.

Redundant Air System. An independent secondary underwater breathing system (i.e., a pony bottle with first and second stage, or a pony bottle supplying a bailout block). NFPA 1670, 2004 ed.

Reference Plane. The plane that is 102.5 mm down from the top of the head and parallel to the basic plane on an ISO size J headform. NFPA 1971, 2000 ed.

Referral. An act or process by which an individual and/or family gain access to a program or community resources. NFPA 1035, 2005 ed.

Refinery. A plant in which flammable or combustible liquids are produced on a commercial scale from crude petroleum, natural gasoline, or other hydrocarbon sources. NFPA 30, 2003 ed.

Region. A geographic area that includes the local and neighboring jurisdiction for an EMS agency. NFPA 473, 2002 ed.

Registered Professional Engineer. A person who is registered as a professional engineer in the state where the work is to be performed. NFPA 1670, 2004 ed.

Regulated Waste. Liquid or semi-liquid blood, body fluids, or other potentially infectious materials; contaminated items that would release blood, body fluids, or other potentially infectious materials in a liquid or semi-liquid state if compressed; items that are caked with dried blood, body fluids, or other potentially infectious materials and are capable of releasing these materials during handling; contaminated sharps; and pathological and microbiological wastes containing blood, body fluids, or other potentially infectious materials. NFPA 1581, 2005 ed.

Rehabilitation. The process of providing rest, rehydration, nourishment, and medical evaluation to members who are involved in extended or extreme incident scene operations. (See Figure R.3.) NFPA 1584, 2003 ed.

Figure R.3 Fire Fighter Rehabilitation.

Reinforcement. *As applied to maintenance of fire fighting protective ensembles:* An additional layer placed in or on an element. NFPA 1851, 2001 ed. *As applied to fire hose:* The structural support for fire hose that is often in the form of woven yarn. NFPA 1961, 2002 ed.

Rekindle. A return to flaming combustion after apparent but incomplete extinguishment. NFPA 921, 2004 ed.

Related Duties. Any and all functions that fire department members can be called upon to perform in the execution of their duties. NFPA 1710, 2004 ed.

Relay-Supply Hose. A single-jacket fire hose of $3\frac{1}{2}$-in. (90-mm) diameter or larger used to move large volumes of water at low pressure and manufactured prior to January 1987 to meet the requirements of the 1979 and previous editions of NFPA 1961, Standard on Fire Hose. NFPA 1962, 2003 ed.

Relief Valve. A device that allows the bypass of fluids to limit the pressure in a system. NFPA 1914, 2002 ed.

Relocatable Power Tap. A device for indoor use consisting of an attachment plug on one end of a flexible cord and two or more receptacles on the opposite end, and has overcurrent protection. NFPA 1, 2003 ed.

Remote Area. A geographic area that requires a travel distance of at least 8 miles to provide emergency services. NFPA 1720, 2004 ed.

Removable SCBA-Integrated PASS. An SCBA-Integrated PASS that is designed and intended to be readily removed from the PASS/SCBA device to be used independently of the SCBA. NFPA 1982, 1998 ed.

Removable Winch. A winch with quick disconnects for power and controls that can be temporarily mounted on the apparatus at a permanently installed mounting receiver. NFPA 1901, 2003 ed.

Rendezvous Point. A prearranged reference point, that is, road junction, crossroad, or other specified place, where personnel/vehicles responding to an emergency situation initially proceed to receive directions to staging areas or the accident/incident site or both. NFPA 424, 2002 ed.

Repair Garages. Buildings, structures, or portions thereof wherein major repair, painting, or body and fender work is performed on motorized vehicles or automobiles; includes associated floor space used for offices, parking, or showrooms. NFPA 1, 2000 ed.

Repeater. A device for receiving and re-transmitting one-way or two-way communication signals. NFPA 1221, 2002 ed.

Replace. To remove an unserviceable item and install a serviceable counterpart in its place. NFPA 1915, 2000 ed.

Replacement. The removal of an existing component or system and the installation of a similar component or system generally of the same model or the same capability (i.e., "like for like" replacement). NFPA 1912, 2001 ed.

Report. The act of providing an account of facts relating to past events, or acts of an organization or its individuals. NFPA 1401, 2001 ed.

Requisite Equipment. Specific tools and equipment that are critical to performing a specific type of technical rescue. NFPA 1006, 2003 ed.

Requisite Knowledge. Fundamental knowledge one must have in order to perform a specific task. NFPA 1031, 2003 ed.

Requisite Skills. The essential skills one must have in order to perform a specific task. NFPA 1031, 2003 ed.

Rescue. See Rescue Operations.

Rescue Area. Sometimes called the "hot," "danger," or "collapse" zone, an area surrounding the incident site (e.g., collapsed structure or trench) that has a size proportional to the hazards that exist. NFPA 1006, 2003 ed.

Rescue Attendant. A person who is qualified to be stationed outside a confined space to monitor rescue entrants, summon assistance, and perform non-entry rescues. NFPA 1670, 2004 ed.

Rescue Company. A group of fire fighters who work as a unit and are equipped with one or more rescue vehicles. NFPA 1410, 2005 ed.

Rescue Entrant. A person entering a confined space for the specific purpose of rescue. NFPA 1670, 2004 ed.

Rescue Incident. An emergency incident that primarily involves the rescue of persons subject to physical danger and that can include the provision of emergency medical services. NFPA 1500, 2002 ed.

Rescue Mode. A level of operational urgency where there is a chance that a victim will be rescued alive. NFPA 1670, 1999 ed.

Rescue Operations. Those activities directed at locating endangered persons, removing endangered persons from danger, treating the injured at an emergency incident, and providing transport to an appropriate health care facility. NFPA 1951, 2005 ed.

Rescue Service. The confined space rescue team designated by the AHJ to rescue victims from within confined spaces, including operational and technical levels of industrial, municipal, and private sector organizations. *Note: All rescue services meet the following minimum requirements: (a) Each member of the rescue service is provided with, and trained to use properly, the personal* *protective equipment and rescue equipment necessary for making rescues from confined spaces according to his or her designated level of competency. (b) Each member of the rescue service is trained to perform the assigned rescue duties corresponding to his or her designated level of competency. Each member of the rescue service also receives the training required of authorized rescue entrants. (c) Each member of the rescue service practices making confined space rescues by means of simulated rescue operations in which they remove dummies, mannequins, or persons from actual confined spaces or from representative confined spaces. Representative confined spaces should—with respect to opening size, configuration, and accessibility—simulate the types of confined spaces from which rescue is to be performed. (d) Each member of the rescue service is certified to the level of first responder or equivalent according to U.S. Department of Transportation (DOT) First Responder Guidelines. Each member of the rescue service also successfully completes a course in cardiopulmonary resuscitation (CPR) taught through the American Heart Association (AHA) to the level of a "Health Care Provider," through the American Red Cross (ARC) to the "CPR for the Professional Rescuer" level, or through the National Safety Council's equivalent course of study. (e) The rescue service is capable of responding in a timely manner to rescue summons. (f) Each member of the rescue service is properly equipped, trained, and capable of functioning appropriately to perform confined space rescues within the area for which they are responsible at their designated level of competency. This must be confirmed by an annual evaluation of the rescue service's capabilities to verify that the needed capabilities are present to perform confined space rescues in terms of overall timeliness, training, and equipment and to perform safe and effective rescue in those types of spaces to which the team must respond. (g) Each member of the rescue service is aware of the hazards they could confront when called on to perform rescue within confined spaces for which they are responsible. (h) If required to provide confined space rescue within U.S. federally regulated industrial facilities, the rescue service must have access to all confined spaces from which rescue could be necessary so that they can develop appropriate rescue plans and practice rescue operations according to their designated level of competency. NFPA 1670, 1999 ed.*

Rescue Team. A combination of rescue-trained individuals who are equipped and available to respond to and perform technical rescues. (SEE FIGURE R.4.) NFPA 1006, 2003 ed.

Rescue Team Leader. The person designated within the incident command system as rescue group/division officer responsible for direct supervision of the rescue team operations. NFPA 1670, 2004 ed.

FIGURE R.4 Rescue Team.

Rescue Technician. A person who is trained to perform or direct the technical rescue. NFPA 1006, 2003 ed.

Rescue Vehicle. A special vehicle, also known as a heavy rescue or squad, equipped with tools and equipment to perform one or more types of special rescue such as building collapse, confined space, high angle, vehicle extrication, and water rescue. NFPA 1410, 2005 ed.

Reserve Capacity. The ability of a battery to sustain a minimum electrical load in the event of a charging system failure or a prolonged charging system deficit. NFPA 1901, 2003 ed.

Reserve Capacity Rating. The number of minutes a new, fully charged battery at 26.7°C (80°F) can be discharged at 25 amperes while maintaining 1.75 volts per cell or higher. NFPA 414, 2001 ed.

Residential Board and Care Occupancy. A building or portion thereof that is used for lodging and boarding of four or more residents, not related by blood or marriage to the owners or operators, for the purpose of providing personal care services. NFPA 5000, 2002 ed.

Residential Occupancy. An occupancy that provides sleeping accommodations for purposes other than health care or detention and correctional. NFPA 5000, 2002 ed.

Residual Pressure. The pressure that exists in the distribution system, measured at the residual hydrant at the time the flow readings are taken at the flow hydrants. NFPA 1410, 2005 ed.

Resource Assessment. The component of the assessment phase that involves the determination for the need for additional resources. *Note: Resource assessment can be ongoing throughout the entire incident.* NFPA 1670, 2004 ed.

Resources. *As applied to wildland fire fighting:* All personnel and major items of equipment that are available, or potentially available, for assignment to incidents. NFPA 1051, 2002 ed. *As applied to qualifications for public fire and life safety educator:* Any personnel, materials, or both, including volunteer educators, educational or promotional materials, and financial resources, required to meet the needs of a program. NFPA 1035, 2005 ed.

Respirator. A device that provides respiratory protection for the wearer. (See Figure R.5.) NFPA 1994, 2001 ed.

Respiratory Equipment. A positive pressure, self-contained breathing apparatus (SCBA) or combination SCBA/supplied-air breathing apparatus certified by the National Institute for Occupational Safety and Health (NIOSH) and certified as compliant with NFPA 1981, Standard on Open-Circuit Self-Contained Breathing Apparatus for Fire and Emergency Services. NFPA 1991, 2005 ed.

Respiratory Hazard. Any exposure to products of combustion, superheated atmospheres, toxic gases, vapors, or dust, potentially explosive or oxygen-deficient atmospheres, or any condition that creates a hazard to the respiratory system. NFPA 1404, 2002 ed.

Respiratory Protection. Equipment designed to protect the wearer from the inhalation of contaminants. (See Figure R.6.) NFPA 472, 2002 ed.

Respiratory Protection Equipment (RPE). Devices that are designed to protect the respiratory system against exposure to gases, vapors, or particulates. NFPA 1404, 2002 ed.

Respiratory Protection Program. A systematic and comprehensive program of training in the use and maintenance of respiratory protection devices and related equipment. NFPA 1404, 2002 ed.

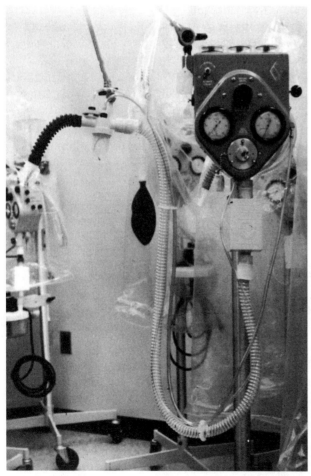

Figure R.5 Respirator.

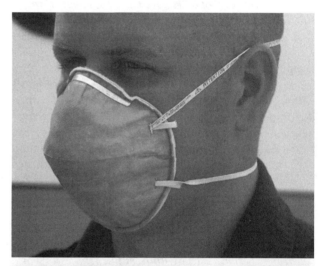
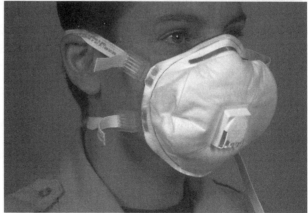

Figure R.6 Two Types of HEPA Respirators.

Responders. Personnel who have responsibility to respond to emergencies such as fire fighters, law enforcement, lifeguards, emergency medical, emergency management, forestry, public health, public works personnel, and other public service personnel. NFPA 1561, 2005 ed.

Responding Personnel. Personnel, whether public or private, available to respond to emergencies. NFPA 1620, 2003 ed.

Response. *As applied to hazardous materials response:* That portion of incident management in which personnel are involved in controlling (defensively or offensively) a hazardous materials incident. NFPA 471, 2002 ed. *As applied to emergency medical services:* The deployment of an emergency service resource to an incident. NFPA 450, 2004 ed.; NFPA 901, 2001 ed.

Response Agency. An organization capable of providing emergency services. NFPA 1670, 1999 ed.

Response Duty. A fire-related service, function, or task identified in the industrial fire brigade organizational statement and assigned to a member to perform. NFPA 600, 2005 ed.

Response Time. *As applied to aircraft:* The total period of time measured from the time of an alarm until the first ARFF vehicle arrives at the scene of an aircraft accident and is in position to apply agent to any fire. NFPA 402, 2002 ed. *As applied to the organization and deployment of fire suppression operations, emergency medical operations, and special operations by fire departments:* The travel time that begins when units are en route to the emergency incident and ends when units arrive at the scene. NFPA 1710, 2004 ed.

Response Unit. A vehicle, equipment, or personnel identified by the AHJ for dispatch purposes. NFPA 1221, 2002 ed.

Responsibility. The accountability of a person or other entity for the event or sequence of events that caused the fire or explosion, spread of the fire, bodily injuries, loss of life, or property damage. NFPA 921, 2004 ed.

Resuscitation Equipment. Respiratory assist devices such as bag-valve masks, oxygen demand valve resuscitators, pocket masks, and other ventilation devices that are designed to provide artificial respiration or assist with ventilation of a patient. NFPA 1581, 2005 ed.

Retention System. The complete assembly by which the helmet is retained in position on the head. NFPA 1971, 2000 ed.

Retirement. The process of permanently removing an element from emergency operations service in the organization. NFPA 1851, 2001 ed.

Retrieval System. Combinations of rescue equipment used for nonentry (external) rescue of persons from confined spaces. NFPA 1670, 1999 ed.

Retroreflection. The reflection of light in which the reflected rays are preferentially returned in the direction close to the opposite of the direction of the incident rays, with this property being maintained over wide variations of the direction of the incident rays. NFPA 1976, 2000 ed.

Retroreflective Markings. A material that reflects and returns a relatively high proportion of light in a direction close to the direction from which it came. NFPA 1976, 2000 ed.

RIC. See Rapid Intervention Crew/Company.

Rigging. The process of building a system to move or stabilize a load. NFPA 1006, 2003 ed.

Rigging Systems. Systems used to move people or loads that can be configured with rope, wire rope, or cable and utilize different means, both mechanical and manual, to move the load. NFPA 1006, 2003 ed.

Ring. An auxiliary equipment system component; an ungated load-bearing connector. NFPA 1983, 2001 ed.

Riser. A pipe leading from the fire main to fire station (hydrants) on upper deck levels. NFPA 1405, 2001 ed.

Risk. A measure of the probability and severity of adverse effects that result from an exposure to a hazard. NFPA 1451, 2002 ed.

Risk Assessment. An assessment of the likelihood, vulnerability, and magnitude of incidents that could result from exposure to hazards. NFPA 1250, 2004 ed.

Risk Control. The management of risk through stopping losses via exposure avoidance, prevention of loss (addressing frequency) and reduction of loss (addressing severity), segregation of exposures, and contractual transfer techniques. NFPA 1250, 2004 ed.

Risk Financing. The aspect of risk management that provides ways to pay for losses. NFPA 1250, 2004 ed.

Risk Management. The process of planning, organizing, directing, and controlling the resources and activities of an organization in order to minimize detrimental effects on that organization. NFPA 1250, 2004 ed.

Risk/Benefit Analysis. A decision made by a responder based on a hazard identification and situation assessment that weighs the risks likely to be taken against the benefits to be gained for taking those risks. NFPA 1670, 2004 ed.

Road. Any accessway, not including a driveway, that gives access to more than one parcel and is primarily intended for vehicular access. NFPA 1144, 2002 ed.

Road Spray Location. Any underbody or underchassis location that is subject to road spray. NFPA 1901, 2003 ed.

Room. The space or area bounded by walls. NFPA 901, 2001 ed.

Rope. A compact but flexible, torsionally balanced, continuous structure of fibers produced from strands that are twisted, plaited, or braided together, and that serve primarily to support a load or transmit a force from the point of origin to the point of application. NFPA 1983, 2001 ed.

Rope Grab Device. An auxiliary equipment system component; a device used to grasp a life safety rope for the purpose of supporting loads; can be used in ascending a fixed line. NFPA 1983, 2001 ed.

Rope Rescue Equipment. Components used to build rope rescue systems including life safety rope, life safety harnesses, and auxiliary rope rescue equipment. (See Figure R.7.) NFPA 1670, 2004 ed.

Rope Rescue System. A system comprised of rope rescue equipment and an appropriate anchor system intended for use in the rescue of a subject. NFPA 1670, 2004 ed.

Rope-Based Mechanical Advantage System. A rope rescue system component incorporating the reeving of rope through moving pulleys (or similar devices) to create mechanical advantage. NFPA 1670, 2004 ed.

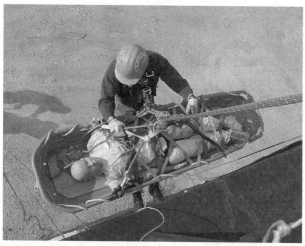

Figure R.7 Rope Rescue Equipment.

Rotation Gear. The main gear of an aerial device that is used for the rotation of the turntable. NFPA 1914, 2002 ed.

Rotation Gear Reduction Box. The mechanism of an aerial device that transfers hydraulic or electric power to the rotation gear, creating the torque necessary to rotate the turntable. NFPA 1914, 2002 ed.

Rotation Lock. A strong friction or other positive-locking device (e.g., holding valve) that retains the turntable in any desired position. NFPA 1914, 2002 ed.

Rotational-Type Control. A control that rotates in a plane perpendicular to the axis of the nozzle. NFPA 1964, 2003 ed.

Routine Cleaning. The light cleaning of ensembles or elements performed by the end user without taking the elements out of service. NFPA 1851, 2001 ed.

rpm. Revolutions per minute. NFPA 1925, 2004 ed.

Rubber-Gasketed Fitting. A device for providing a leakproof connection between two pieces of pipe while allowing moderate movement of one pipe relative to the other. NFPA 414, 2001 ed.

Rung Cap Casting. A casting that can be riveted to the outside of the base rail over the ends of each rung on an aerial ladder. NFPA 1914, 2002 ed.

Rungs. The ladder crosspieces on which a person steps while ascending or descending. NFPA 1931, 2004 ed.

Runoff. Liquids that flow by gravity away from an aircraft accident and might include aviation fuel (ignited or not), water from fire fighting streams, liquid cargo, or a combination of these liquids. NFPA 402, 2002 ed.

Runway. A defined rectangular area on a land airport prepared for the landing and taking off of aircraft along its length. *Note: Runways are normally numbered relative to their magnetic direction.* NFPA 402, 2002 ed.

Rural. Any area wherein residences and other developments are scattered and intermingled with forest, range, or farmland and native vegetation or cultivated crops. NFPA 1143, 2003 ed.

Rural Area. As defined by the U.S. Census Bureau, an area with fewer than 500 people per square mile. NFPA 1720, 2004 ed.

Sacrificial Anode System. Galvanic corrosion protection that employs zinc, aluminum, or magnesium anodes connected to the vessel's hull. The anodes dissolve away over time. NFPA 1925, 2004 ed.

Safe Zone. In a trench, the area that projects 0.61 m (2 ft) in all directions around an installed cross brace or wale that is a component of an existing approved shoring system. NFPA 1006, 2003 ed.

Safely. To perform the assigned tasks without injury to self or others, to the environment, or to property. NFPA 472, 2002 ed.

Safetied (Safety Knot). A securement of loose rope end issuing from a completed knot, usually fashioned by tying the loose end around another section of rope to form a knot; the means by which the loose end is prevented from slipping through the primary knot. (See Figure S.1.) NFPA 1006, 2003 ed.

Safety Alert. The action by which a manufacturer identifies a specific compliant product or a compliant product component, provides notice to users of the compliant product, and informs the marketplace and distributors of potential safety concerns regarding the product or component. NFPA 1999, 2003 ed.

Safety Diver. An on-site diver available in a sufficient state of readiness to assist another diver in the water. NFPA 1670, 1999 ed.

Safety Glasses. An eye and face protection device intended to help protect the wearer's eyes. NFPA 1951, 2005 ed.

Safety Officer. An individual appointed by the authority having jurisdiction as qualified to maintain a safe working environment at all live fire training evolutions. NFPA 1403, 2002 ed.

Safety Specialist. An individual who has the expertise, knowledge, and professional experience to achieve control or reduction of occupational hazards and exposures. Note: *The safety specialist is assigned to assist the health and safety officer as part of the safety staff.* NFPA 1521, 2002 ed.

Safety Stop Mechanism. A device that is located on the aerial device and prevents raising the elevating platform booms or sections beyond safe, operating, horizontal or vertical angles. NFPA 1914, 2002 ed.

Safety Unit. A member or members assigned to assist the incident safety officer. The tactical level management unit that can be comprised of the incident safety officer alone or with additional assistant safety officers assigned to assist in providing the level of safety supervision appropriate for the magnitude of the incident and the associated hazards. NFPA 1521, 2002 ed.

Safety Valve. An automatic oil control valve of the "on" and "off" type (without any bypass to the burner) that is actuated by a safety control or by an emergency device. NFPA 31, 2001 ed.

Sagging. Straining of the ship that tends to make the middle portion lower than the bow and stern. NFPA 1405, 2001 ed.

Sail Area. The area of the ship that is above the waterline and that is subject to the effects of wind, particularly a crosswind on the broad side of a ship. NFPA 1405, 2001 ed.

Sales Display Area. The area of a mercantile occupancy that is open to the public for the purpose of viewing and purchasing goods, wares, and merchandise. Individuals are free to circulate among the items, which are typically displayed on shelves, racks, or on the floor. NFPA 30B, 2002 ed.

Figure S.1 Safety Knot.

Sally Port (Security Vestibule). A compartment provided with two or more doors where the intended purpose is to prevent continuous and unobstructed passage by allowing the release of only one door at a time. NFPA 101, 2003 ed.

Salvage. The restoration of a distressed vessel to normal condition, usually the removal of water from inside the hull. NFPA 1925, 2004 ed.

Salvage Vehicle. A vehicle that is dismantled for parts or awaiting destruction. NFPA 1, 2003 ed.

Sample. (1) The ensemble, element, item, or component that is conditioned for testing. (2) Ensembles, elements, items, or components that are randomly selected from the manufacturer's production line, from the manufacturer's inventory, or from the open market. NFPA 1977, 2005 ed.

Sample Specimen. A specified number of units of the single layer of material or the composite of the materials, or a specified number of units of the components, both of which have been taken from a manufacturer's current production lot used in actual construction of station/work uniforms proposed for being labeled as compliant with the requirements of this standard. NFPA 1975, 1999 ed.

Sampling. The process of collecting a representative amount of gas, liquid, or solid for analytical purposes. NFPA 471, 2002 ed.

Sanitary Sewer. A sewer that carries liquid and water-carried wastes from residences, commercial buildings, industrial plants, and institutions together with minor quantities of storm water, surface water, and groundwater that are not admitted intentionally. NFPA 820, 2003 ed.

Sanitize. The removal of dirt and the inhibiting of the action of agents that cause infection or disease. NFPA 1404, 2002 ed.

SAR. See Supplied Air Respirator.

Satellite Building. A structure that can be adjacent to but separated from the airport terminal building, accessible above ground or through subway passages, and used to provide flight service operations, such as passenger check-in, waiting rooms, food service, enplaning or deplaning, etc. NFPA 1, 2003 ed.

Scavenged Gas. A residual process gas that is collected for treatment or release at a location remote from the site of use. NFPA 1, 2003 ed.

SCBA. See Self-Contained Breathing Apparatus.

SCBA Fill Hose. See Fill Hose.

SCBA Fill Station. A containment enclosure for refilling self-contained breathing cylinders to guard personnel from fragments due to accidental cylinder rupture. NFPA 1901, 2003 ed.

SCBA-Integrated PASS. A removable or nonremovable PASS that is an integral part of a PASS/SCBA device. NFPA 1982, 1998 ed.

Scene Security. The means used to prevent or restrict entry to the scene of a rescue incident, either during or following the emergency. NFPA 1006, 2003 ed.

Scientific Method. The systematic pursuit of knowledge involving the recognition and formulation of a problem, the collection of data through observation and experiment, and the formulation and testing of a hypothesis. NFPA 921, 2004 ed.

Scorch. Discoloring (browning or blackening) of a material, a characteristic of the overheat condition. NFPA 901, 2001 ed.

Screw Jack. Shoring system component made of sections of threaded bar stock that are incorporated with lengths of pipe or wood. NFPA 1006, 2003 ed.

Screw Thread Coupling or Adapter. A coupling or adapter in which the mating is achieved with the use of threads. (See Figure S.2.) NFPA 1963, 1998 ed.

Scrubbing. A process of agitating foam solution and air in a confined space such as a hose, pipe, or mixing chamber to produce bubbles. NFPA 1145, 2000 ed.

SCUBA. Self-contained underwater breathing apparatus. NFPA 1006, 2003 ed.

Scupper. An opening in the side of a vessel through which rain, sea, or fire fighting water is discharged. NFPA 1405, 2001 ed.

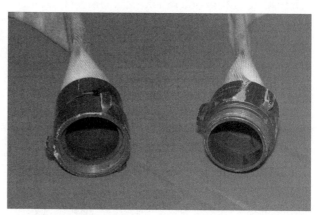

Figure S.2 Threaded Couplings.

Seam. Any permanent attachment of two or more chemical-protective clothing materials, excluding external fittings, gaskets, and suit closure assemblies, in a line formed by joining the separate material pieces. NFPA 1991, 2005 ed.

Seam Assembly. The structure obtained when fabrics are joined by means of a seam. NFPA 1971, 2000 ed.

Seams (Major). Those seam assemblies where rupture exposes the wearer to immediate danger. NFPA 1977, 2005 ed.

Seams (Major A). Outer shell layer seam assemblies where rupture could reduce the protection of the garment by exposing the inner layers such as the moisture barrier, the thermal barrier, the wearer's station/work uniform, other clothing, or skin. NFPA 1971, 2000 ed.

Seams (Major B). Moisture barrier or thermal barrier seam assemblies where rupture could reduce the protection of the garment by exposing the next layer of the garment, the wearer's station/work uniform, other clothing, or skin. NFPA 1971, 2000 ed.

Seams (Minor). Remaining seam assemblies that are not classified as Major A or Major B seams. NFPA 1971, 2000 ed.

Search Function. General area search, reconnaissance, victim location identification, and hazard identification or flagging. NFPA 1006, 2003 ed.

Search Marking System. A separate and distinct marking system used to identify information related to the location of a victim(s). NFPA 1670, 2004 ed.

Search Operations. Any land-based operations involving the search for victims or body recovery. NFPA 1951, 2005 ed.

Search Parameters. The defined search area and scope. NFPA 1006, 2003 ed.

Seasonal Needs. The periodic operating contingencies of a particular geographic region that are established to overcome a specific set of circumstances (e.g., crop harvest, rainy season). NFPA 1401, 2001 ed.

Seat Belt. A two-point lap belt, a three-point lap/shoulder belt, or a four-point lap/shoulder harness for vehicle occupants designed to limit their movement in the event of an accident, rapid acceleration, or rapid deceleration by securing individuals safely to a vehicle in a seated position. NFPA 1500, 2002 ed.

Seat Circumference. Lower torso garment measurement from 25 mm (1 in.) above bottom of fly curve from folded edge to folded edge, multiplied by 2 to obtain circumference. NFPA 1977, 2005 ed.

Seat of Explosion. A craterlike indentation created at the point of origin of an explosion. NFPA 921, 2004 ed.

Seated Explosion. An explosion with a highly localized point of origin, such as a crater. NFPA 921, 2004 ed.

Seaworthy. A vessel's capability to perform its mission in adverse sea or weather conditions. NFPA 1925, 2004 ed.

Secondary Access. Openings created by rescuers that provide a pathway to trapped and/or injured victims. NFPA 1670, 2004 ed.

Secondary Collapse. Causes or conditions that could contribute to a subsequent collapse in a building. NFPA 1006, 2003 ed.

Secondary Combustion Air. The air externally supplied to the flame in the combustion zone. NFPA 97, 2003 ed.

Secondary Containment Tank. A tank that has an inner and outer wall with an interstitial space (annulus) between the walls and that has a means for monitoring the interstitial space for a leak. NFPA 30, 2003 ed.

Secondary Contamination. The process by which a contaminant is carried out of the hot zone and contaminates people, animals, the environment, or equipment. NFPA 472, 2002 ed.

Secondary (Design) Water Supply. The estimated rate of flow [expressed in gpm (L/min) for a prescribed time period] that is necessary to control a major fire in a building or structure. NFPA 1142, 2001 ed.

Secondary Explosion. Any subsequent explosion resulting from an initial explosion. NFPA 921, 2004 ed.

Secondary Storage Facility. Warehouse facilities established to house books and other collections. NFPA 909, 2005 ed.

Sector. Either a geographic or functional assignment. NFPA 1561, 2005 ed.

Security Systems. Several items of equipment, processes, design features and actions or behaviors intended to discover, report, deter, or delay criminal acts from being perpetrated against persons or property. NFPA 1031, 2003 ed.

Security Vestibule. A compartment that is provided with two or more doors to prevent continuous and unobstructed passage by allowing the release of only one door at a time. NFPA 1221, 2002 ed.

Segregated Storage. Storage located in the same room or inside area that is physically separated by distance from incompatible materials. NFPA 1, 2003 ed.

Selection. The process of determining what protective clothing and equipment is necessary for protection of fire and emergency service responders from an anticipated, specific hazard, or other activity, the procurement of the appropriate protective clothing and equipment, and the choice of the proper protective clothing and equipment for a specific hazard or activity at an emergency scene. NFPA 1851, 2001 ed.

Self-Closing. Equipped with an approved device that ensures closing after opening. NFPA 101, 2003 ed.

Self-Contained Breathing Apparatus (SCBA). An atmosphere-supplying respirator that supplies a respirable air atmosphere to the user from a breathing air source that is independent of the ambient environment and designed to be carried by the user. (See Figure S.3.) *Note: Where the term self-contained breathing apparatus is used without any qualifier, it indicates only open-circuit self-contained breathing apparatus or combination SCBA/SARs. Combination SCBA/SARs are encompassed by the terms self-contained breathing apparatus or SCBA.* NFPA 1981, 2002 ed.

Self-Destructive Action. Interaction of materials in a manner that leads to deterioration. NFPA 1983, 2001 ed.

Self-Heating. The result of exothermic reactions, occurring spontaneously in some materials under certain conditions, whereby heat is generated at a rate sufficient to raise the temperature of the material. NFPA 921, 2004 ed.

Self-Ignition. Ignition resulting from self-heating. *Note: The term self-ignition is synonymous with spontaneous ignition.* NFPA 921, 2003 ed.

Self-Ignition Temperature. The minimum temperature at which the self-heating properties of a material lead to ignition. NFPA 921, 2004 ed.

Self-Rescue. Escaping or exiting a hazardous area under one's own power. NFPA 1006, 2003 ed.

Sensing. A functional mode in which the PASS motion detector is activated and is monitoring the motion of the wearer, and causes pre-alarm signal and subsequent transfer to the alarm signal after not detecting motion after a specified period of time. NFPA 1982, 1998 ed.

Separate. A material response evidenced by splitting or delaminating. NFPA 1971, 2000 ed.

Separation. An intervening space (as opposed to barrier). NFPA 550, 2002 ed.

Separation of Hazards. Physically separated by a specified distance, construction, or appliance. NFPA 55, 2005 ed.

Service Brake. A system capable of decelerating the vehicle at a controlled rate to a desired reduced speed or complete stop. NFPA 414, 2001 ed.

Service Life. The period for which a certified product is useful before retirement. NFPA 1981, 2002 ed.

Service Request. Any communication from the public or other agency that prompts action by a telecommunicator. NFPA 1061, 2002 ed.

Service Station, Located Inside a Building. That portion of an automotive service station located within the perimeter of a building or building structure that also contains other occupancies. *Note: The service station is permitted to be enclosed or partially enclosed by the building walls, floors, ceilings, or partitions or it may be open to the outside. The service station dispensing area is that area of the service station required for dispensing of fuels to motor vehicles. Dispensing of fuel at manufacturing, assembly, and testing operations is not included within this definition.* NFPA 1, 2000 ed.

Service Test. Hydrostatic test conducted by users on all in-service hose to determine suitability for continued service. NFPA 1962, 2003 ed.

Service Test Pressure. A pressure equal to approximately 110 percent of the operating pressure. NFPA 1961, 2002 ed.

Service Testing. The regular, periodic inspection and testing of apparatus and equipment, according to an established schedule and guideline, to ensure that they are in safe and functional operating condition. NFPA 1500, 2002 ed.

Figure S.3 Open-Circuit SCBA.

SETIQ. The Emergency Transportation System for the Chemical Industry in Mexico. NFPA 472, 2002 ed.

Severe Service. Those conditions that apply to the rigorous, harsh, and unique applications of fire apparatus, including but not limited to local operating and driving conditions, frequency of use, and manufacturer's severe service (duty) parameters. NFPA 1915, 2000 ed.

Sewn Seam. A series of stitches joining two or more separate plies of material(s) of planar structure, such as textiles. NFPA 1975, 2004 ed.

Sewn Seam Strength. The maximum resistance to rupture of the junction formed by stitching together two or more planar structures, such as textile fabrics. NFPA 1977, 2005 ed.

Shaft Alley. A narrow, watertight compartment through which the propeller shaft passes from the aft engine room bulkhead to the propeller. NFPA 1405, 2001 ed.

Shaftway. A tunnel or alleyway through which the drive shaft or rudder shaft passes. NFPA 1405, 2001 ed.

Shall. Indicates a mandatory requirement. NFPA Official Definition.

Shank. Reinforcement to the area of protective footwear designed to provide additional support to the instep. NFPA 1971, 2000 ed.

Sharps Container. A container that is closable, puncture-resistant, disposable, and leakproof on the sides and bottom; red in color or displays the universal biohazard symbol; and designed to store sharp objects after use. NFPA 1581, 2005 ed.

Sheer. Upper edge of hull exterior at the intersection with the deck. NFPA 1925, 2004 ed.

Sheeting. The members of a shoring system that support the sides of an excavation and are in turn supported by other members of the shoring system. NFPA 1670, 2004 ed.

Sheeting or Sheathing. A component of a shoring system with a large surface area supported by the uprights and cross bracing of the shoring system that is used to retain the earth in position when loose or running soils are encountered. NFPA 1006, 2003 ed.

Shell. *As applied to wildland fire fighting:* A helmet without the suspension system, accessories, and fittings. NFPA 1977, 2005 ed. *As applied to fire fighting:* The outermost layer of the protective ensemble element composite. NFPA 1976, 2000 ed.

Shield (or Shield System). A structure that is able to withstand the forces imposed on it by a cave-in and thereby protect employees within the structures. NFPA 1670, 2004 ed.

Shop Drawings. Scaled working drawings, equipment cutsheets, and design calculations. NFPA 1031, 2003 ed.

Shore-Based Rescue. Any technique or procedure that provides a means for extracting a person from the water that does not require any member of the rescue team to leave the safety of the shore. NFPA 1006, 2003 ed.

Shoring (or Shoring System). A structure such as a metal hydraulic, pneumatic/mechanical, or timber system that supports the sides of an excavation and is designed to prevent cave-ins. NFPA 1670, 2004 ed.

Shoring Team. The group of individuals, with established communications and leadership, assigned to construct, move, place, and manage the shoring or shoring system inside the space, trench, or excavation. NFPA 1670, 2004 ed.

Short Circuit. An abnormal connection of low resistance between normal circuit conductors where the resistance is normally much greater; this is an overcurrent situation but it is not an overload. NFPA 921, 2004 ed.

Short-Term Exposure Limit (STEL). The concentration to which it is believed that workers can be exposed continuously for a short period of time without suffering from irritation, chronic or irreversible tissue damage, or narcosis of a degree sufficient to increase the likelihood of accidental injury, impairment of self-rescue, or the material reduction of work efficiency, without exceeding the daily permissible exposure limit (PEL). NFPA 55, 2005 ed.

Should. Indicates a recommendation or that which is advised but not required. NFPA Official Definition.

Side Slope. An angle measured as either the percent of slope or the tilt angle at which a vehicle would become unstable should the vehicle be placed on the side of a steep, angled hill or sloped surface. NFPA 414, 2001 ed.

Sign. A visual indication whether in pictorial or word format that provides a warning to the operator or other persons near the apparatus. NFPA 1901, 2003 ed.

Signaling Device. Any resource that provides a distinct and predictable display, noise, or sensation that can be used to communicate a predetermined message or to attract the attention of other persons as desired by the initiator of the signal. NFPA 1006, 2003 ed.

Simple Asphyxiant Gas. A gas that does not provide sufficient oxygen to support life and that has none of the other physical or health hazards. NFPA 1, 2003 ed.

Simple Rope Mechanical Advantage System. A rope mechanical advantage system containing the following: (a) A single rope; (b) One or more moving pulleys (or similar devices), all traveling at the same speed and in the same direction, attached directly or indirectly to the load; (c) In the case of mechanical advantage systems greater than 2:1, one or more stationary pulleys or similar devices. NFPA 1670, 1999 ed.

Simplex Radio Channel. A radio channel using a singular frequency that allows transmission or reception only at a given time. (See Figure S.4.) NFPA 1221, 2002 ed.

Simulation. The repeatable act of carrying out a job performance requirement in a safe environment that reproduces actual job performance conditions to the fullest possible extent. NFPA 1000, 2000 ed.

Single Fire Apparatus. A vehicle on a single chassis frame. NFPA 1915, 2000 ed.

Single Jacket. A construction consisting of one woven jacket. NFPA 1962, 2003 ed.

Single Ladder. A non-self-supporting ground ladder, nonadjustable in length, consisting of only one section. NFPA 1931, 2004 ed.

Single-Point Anchor System. An anchor system configuration utilizing a single anchor point to provide the primary support for the rope rescue system. A single-point anchor system includes those anchor systems that utilize one or more additional nonloaded anchor points as backup to the primary anchor point. NFPA 1670, 2004 ed.

Single-Row Racks. Racks that have no longitudinal flue space and that have a width up to 1.8 m (6 ft) with aisles at least 1.1 m (3.5 ft) from other storage. NFPA 13, 2002 ed.

Single-Use Item. Items that are designed to be used one time and then disposed of. NFPA 1999, 2003 ed.

Site. The entire premises within the governed property lines that contains one or more facilities. NFPA 1081, 2001 ed.

Site Operations. The activities to be undertaken at a specific site to manage the rescue efforts. NFPA 1006, 2003 ed.

Site Stabilization. Those activities directed at mitigating the dangerous elements of an emergency incident. NFPA 1951, 2005 ed.

Site-Specific Hazard. A hazard that is present at the specific facility for which the industrial fire brigade has been organized. NFPA 600, 2005 ed.

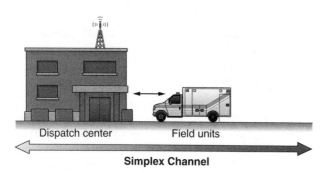

Figure S.4 Simplex Radio Channel Transmission.

Situation Analysis. The process of evaluating the severity and consequences of an incident and communicating the results. NFPA 1600, 2004 ed.

Size Up. The observation and evaluation of existing factors that are used to develop objectives, strategy, and tactics for fire suppression. NFPA 1051, 2002 ed.

Size-Up (Risk Assessment). A mental process of evaluating the influencing factors at an incident prior to committing resources to a course of action. NFPA 405, 2004 ed.

Skills. Behaviors or actions that require practice in order to be performed satisfactorily. *Note: The skills or abilities can be manual, social, interpersonal, or intellectual.* NFPA 405, 2004 ed.

Skin. The outer covering of an aircraft fuselage, wings, and empennage. NFPA 402, 2002 ed.

Sleeve Length. Upper torso garment measurement from center back at bottom of collar seam diagonally across back and down sleeve to bottom edge of cuff. In other specified instances, it is a measurement from center sleeve setting seam at shoulder to bottom edge of sleeve. NFPA 1977, 2005 ed.

Slide Blocks. Blocks made of a variety of materials (e.g., brass, nylatron) that act as spacing devices, wear strips, or wear pads. NFPA 1914, 2002 ed.

Slip-On Fire Fighting Module. A self-contained unit that includes an auxiliary engine-driven pump, piping, a tank, and hose storage that is designed to be placed on a truck chassis, utility bed, flatbed, or trailer. NFPA 1906, 2001 ed.

Slope. Upward or downward incline or slant, usually calculated as a percentage. NFPA 1144, 2002 ed.

Slope of Grain Wood Irregularities. A deviation of the fiber direction from a line parallel to the sides of the piece. NFPA 1931, 2004 ed.

Sloping System. A protecting system that uses inclined excavating to form sides that are inclined away from the excavation so as to prevent cave-in. NFPA 1670, 1999 ed.

Slough-In. A type of collapse characterized by an interior portion of the trench wall spalling out and potentially leaving an overhanging ledge or void that needs to be filled. NFPA 1006, 2003 ed.

Slow-Operating Valve. A valve that has a mechanism to prevent movement of the flow-regulating element from the fully closed position to the fully opened position or vice versa in less than 3 seconds. NFPA 1901, 2003 ed.

Slug Flow. The discharge of distinct pockets of air and water or weak foam solution due to the insufficient or uneven mixing of foam concentrate, water, and air in a compressed air foam system (CAFS). NFPA 1145, 2000 ed.

Smoke. The airborne solid and liquid particulates and gases evolved when a material undergoes pyrolysis or combustion, together with the quantity of air that is entrained or otherwise mixed into the mass. NFPA 318, 2002 ed.

Smoke Barrier. A continuous membrane, or a membrane with discontinuities created by protected openings, where such membrane is designed and constructed to restrict the movement of smoke. NFPA 5000, 2002 ed.

Smoke Compartment. A space within a building enclosed by smoke barriers on all sides, including the top and bottom. NFPA 101, 2003 ed.

Smoke Condensate. The condensed residue of suspended vapors and liquid products of incomplete combustion. NFPA 921, 2004 ed.

Smoke Detector. A device that detects visible or invisible particles of combustion. NFPA 72, 2002 ed.

Smoke Ejector. A mechanical device, similar to a large fan, that can be used to force heat, smoke, and gases from a post-fire environment and draw in fresh air. (See Figure S.5.) NFPA 402, 2002 ed.

Smoke Partition. A continuous membrane that is designed to form a barrier to limit the transfer of smoke. NFPA 101, 2003 ed.

Smoking. The use or carrying of a lighted pipe, cigar, cigarette, tobacco, or any other type of smoking substance. NFPA 1, 2003 ed.

Figure S.5 Smoke Ejector.

Smoking Area. A designated area where smoking is permitted within a premises where smoking is otherwise generally prohibited. NFPA 1, 2003 ed.

Smoldering. Combustion without flame, usually with incandescence and smoke. NFPA 921, 2004 ed.

Snap-Link. An auxiliary equipment system component; a self-closing, gated, load-bearing connector. NFPA 1983, 2001 ed.

Software. *As applied to fire service life safety rope:* A type of auxiliary equipment that includes, but is not limited to, anchor straps, pick-off straps, and rigging slings. NFPA 1983, 2001 ed. *As applied to technical search and rescue incidents:* A flexible fabric component of rope rescue equipment that can include, but is not limited to, anchor straps, pick-off straps, and rigging slings. NFPA 1670, 2004 ed.

SOLAS. The International Convention for the Safety of Life at Sea, 1974. NFPA 1405, 2001 ed.

Soldier Shoring or Skip Shoring. A shoring system that employs a series of uprights spaced at intervals with the exposed soil of the trench wall showing. NFPA 1006, 2003 ed.

Solidification. The process whereby a hazardous liquid is treated chemically so that solid material results. NFPA 471, 2002 ed.

Soot. Black particles of carbon produced in a flame. NFPA 921, 2004 ed.

Source Individual. Any individual, living or dead, whose blood, body fluids, or other potentially infectious materials has been a source of occupational exposure to a member. NFPA 1581, 2005 ed.

Spalling. Chipping or pitting of concrete or masonry surfaces. NFPA 921, 2004 ed.

Span of Control. The maximum number of personnel or activities that can be effectively controlled by one individual (usually three to seven). NFPA 1006, 2003 ed.

Spark. A moving particle of solid material that emits radiant energy due either to its temperature or the process of combustion on its surface. NFPA 654, 2000 ed.

Special Operations. Those emergency incidents to which the fire department responds that require specific and advanced training and specialized tools and equipment. (See Figure S.6.) NFPA 1500, 2002 ed.

Special Purpose Fire Fighting Vessel. Any vessel built for another purpose but provided with fixed fire fighting capabilities (e.g., fire tug, work boat, yard patrol boat, hovercraft). NFPA 1925, 2004 ed.

Special Services Fire Apparatus. A multipurpose vehicle that primarily provides support services at emergency scenes. NFPA 1901, 2003 ed.

FIGURE S.6 Specialists Trained in Rope Rescue.

Special Use. A use that includes, but is not limited to, events or occurrences during which life safety–threatening situations or fire hazards exist or are likely to exist as determined by the AHJ. NFPA 1, 2002 ed.

Specialized Apparatus. A fire department emergency vehicle that provides support services at emergency scenes, including command vehicles, rescue vehicles, hazardous material containment vehicles, air supply vehicles, electrical generation and lighting vehicles, or vehicles used to transport equipment and personnel. NFPA 1710, 2004 ed.

Specialized Cleaning. Cleaning to remove hazardous materials or biological agents. NFPA 1851, 2001 ed.

Specialized Equipment. Equipment that is unique to the rescue incident and made available. NFPA 1006, 2003 ed.

Specialized Teams. Emergency response teams with specific skills and equipment that can be needed on the scene. NFPA 1006, 2003 ed.

Specialty Emergency Exercise. One or more specialty agencies fully involved in an exercise to test or give the agency practice in its specialty. NFPA 424, 2002 ed.

Specific Gravity. As applied to gas, the ratio of the weight of a given volume to that of the same volume of air, both measured under the same conditions. NFPA 54, 2002 ed.

Specific Property Use. The purpose for which a specific space, structure, or portion of a structure is used by the owner, tenant, or occupant of the space. NFPA 901, 2001 ed.

Specified Service Life. Time, exposure event, or number of uses to which a compliant product or component is expected to remain functional. NFPA 1852, 2002 ed.

Specimen. The item that undergoes testing; in some cases, the specimen is also the sample. NFPA 1971, 2000 ed.

Spill Prevention Control and Countermeasure (SPCC) Plan. A plan prepared for facilities with a chemical or chemicals that exceed certain capacities in accordance with governmental regulations. NFPA 1620, 2003 ed.

Spiral Reinforcement. A hose reinforcement consisting of pairs of layers of yarn spiraled with no interlacing between the individual layers. The layers of yarn in each pair are spirally wound in opposite directions. A layer of rubber separates each pair of spiraled layers. NFPA 1962, 2003 ed.

Spirit Level. An indicating device that is affixed to a turntable or truck body and is used to verify the levelness of a turntable prior to operating an aerial device. NFPA 1914, 2002 ed.

Splash-Resistant Eyewear. Safety glasses, prescription eyewear with protective side shields, goggles, or chin-length face shields that, when worn properly, provide limited protection against splashes, spray, spatters, or droplets of body fluids. NFPA 1581, 2005 ed.

Split Shaft PTO. A power takeoff (PTO) drive system that is inserted between the chassis transmission and the chassis drive axle and that has the shift mechanism necessary to direct the chassis engine power either to the drive axle or to a fire pump or other accessory. NFPA 1901, 2003 ed.

Split Wood Irregularity. A separation of the wood parallel to the fiber direction due to tearing of the wood fibers. NFPA 1931, 2004 ed.

Spoil Pile (Spoil). A pile of excavated soil next to the excavation or trench. NFPA 1006, 2003 ed.

Spoliation. Loss, destruction, or material alteration of an object or document that is evidence or potential evidence in a legal proceeding by one who has the responsibility for its preservation. NFPA 921, 2004 ed.

Spontaneous Heating. Process whereby a material increases in temperature without drawing heat from its surroundings. NFPA 921, 2004 ed.

Spontaneous Ignition. Initiation of combustion of a material by an internal chemical or biological reaction that has produced sufficient heat to ignite the material. NFPA 921, 2004 ed.

Sports Drink. A fluid replacement beverage that is between 4 percent and 8 percent carbohydrate and contains between 0.5 g and 0.7 g of sodium per liter of solution. NFPA 1584, 2003 ed.

Spray Nozzle. A nozzle with an adjustable pattern and with a control device that shuts off the flow. (SEE FIGURE S.7.) NFPA 1963, 1998 ed.

Spreader. A powered rescue tool that has at least one movable arm that opens to move material. (SEE FIGURE S.8.) NFPA 1936, 2005 ed.

Spreading Force. The force to push or pull that is generated by a spreader rescue tool and that is measured or calculated at the very tips of the spreader arms or ram. NFPA 1936, 2005 ed.

Sprig-up. A line that rises vertically and supplies a single sprinkler. NFPA 13, 2002 ed.

Spring Hinge. A closing device in the form of a hinge with a built-in spring used to hang and close the door. NFPA 80, 1999 ed.

Spring Release Device (Sliding Door, Vertical, Horizontal; Rolling Steel Door). A device that, when activated, releases part of the spring counterbalancing force and causes a door to close. NFPA 80, 1999 ed.

Stabilization. *As applied to airport/community emergency planning:* The medical measures used to restore basic physiologic equilibrium to a patient, to facilitate future definitive care, in order to ensure survival. NFPA 424, 2002 ed. *As applied to hazardous materials incident*

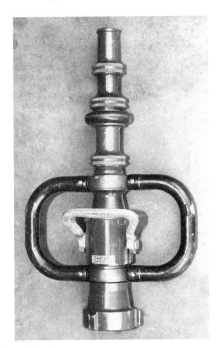

FIGURE S.7 Spray Nozzle.

FIGURE S.8 Hydraulic Spreader.

response: The point in an incident at which the adverse behavior of the hazardous material is controlled. NFPA 471, 2002 ed.

Stabilization Points. Key points where stabilization devices can be installed on a vehicle or machine to keep the vehicle or object from moving during rescue operations. NFPA 1006, 2003 ed.

Stabilizer. A device integral with or separately attached to the chassis of an aerial fire apparatus that is used to increase moments tending to resist overturning the apparatus. NFPA 1914, 2002 ed.

Stabilizer Pad. A plate inserted beneath a stabilizer shoe to give greater surface-bearing area. NFPA 1901, 2003 ed.

Stabilizer Shoe. A permanently mounted shoe on a stabilizer to provide a ground surface area. NFPA 1901, 2003 ed.

Stable Liquid. Any liquid not defined as unstable. NFPA 30, 2003 ed.

Staff Aide. A fire fighter or fire officer assigned to a supervisory chief officer to assist with the logistical, tactical, and accountability functions of incident, division, or sector command. NFPA 1710, 2004 ed.

Staging. A specific function where resources are assembled in an area at or near the incident scene to await instructions or assignments. NFPA 1561, 2005 ed.

Staging Area. A prearranged, strategically placed area, where support response personnel, vehicles, and other equipment can be held in an organized state of readiness for use during an emergency. NFPA 424, 2002 ed.

Stakeholder. An individual, or representative of same, having an interest in the successful completion of a project. NFPA 101, 2003 ed.

Stand-Alone PASS. A PASS that is not an integral part of any other item of protective clothing or protective equipment. NFPA 1982, 1998 ed.

Standard. A document, the main text of which contains only mandatory provisions using the word "shall" to indicate requirements and which is in a form generally suitable for mandatory reference by another standard or code or for adoption into law. Nonmandatory provisions shall be located in an appendix or annex, footnote, or fine-print note and are not to be considered a part of the requirements of a standard. NFPA Official Definition.

Standard Cubic Feet per Minute (scfm). An expression of airflow rate in which the airflow rate is corrected to standard temperature and pressure. NFPA 1901, 2003 ed.

Standard Cubic Foot (scf). One cubic foot of gas at 21°C (70°F) and an absolute pressure of 101.325 kPa (14.7 psia). NFPA 52, 2002 ed.

Standard Deviation. A parameter that indicates the way in which a probability function is centered around its mean. NFPA 1983, 2001 ed.

Standard Equipment and Tools. Investigator's tools and equipment that every investigator must carry. NFPA 1033, 2003 ed.

Standard Operating Guideline. An organizational directive that establishes a course of action or policy. NFPA 1670, 2004 ed.

Standard Operating Procedure (SOP). A written organizational directive that establishes or prescribes specific operational or administrative methods to be followed routinely for the performance of designated operations or actions. NFPA 1521, 2002 ed.

Standard Temperature and Pressure STP. A temperature of 21°C (70°F) and a pressure of 1 atmosphere (14.7 psi or 760 mm Hg). NFPA 1, 2003 ed.

Standing Orders. A direction or instruction for delivering patient care without on-line medical oversight backed by authority of the system medical director. NFPA 450, 2004 ed.

Standpipe System. An arrangement of piping, valves, hose connections, and allied equipment installed in a building or structure, with the hose connections located in such a manner that water can be discharged in streams or spray patterns through attached hose and nozzles, for the purpose of extinguishing a fire, thereby protecting a building or structure and its contents in addition to protecting the occupants. This is accomplished by means of connections to water supply systems or by means of pumps, tanks, and other equipment necessary to provide an adequate supply of water to the hose connections. See Figure S.9, a Class I standpipe system that provides water for the fire department hose lines, and Figure S.10, a Class II standpipe, which is meant to be used by building occupants. NFPA 14, 2003 ed.

Figure S.9 Class I Standpipe System.

Station/Work Uniform Garment. Textile apparel that covers the torso and limbs or parts of limbs, excluding heads, hands, and feet. NFPA 1975, 2004 ed.

Station/Work Uniforms. Nonprimary protective garments that are intended to be worn by fire and emergency services personnel while on duty. NFPA 1975, 2004 ed.

Figure S.10 Class II Standpipe System.

Stationary Tank. A packaging designed primarily for stationary installations not intended for loading, unloading, or attachment to a transport vehicle as part of its normal operation in the process of use. NFPA 55, 2005 ed.

Staypoles (Tormentors). Poles attached to each beam of the base section of extension ladders and used to assist in raising the ladder and to help provide stability of the raised ladder. (See Figure S.11.) NFPA 1931, 2004 ed.

Steel Cutting Tools. Hand tools, circular saw, exothermic torch, oxyacetylene torch, and plasma cutter. (See Figure S.12.) NFPA 1006, 2003 ed.

Steering Axle. Any axle designed such that the wheels have the ability to turn the vehicle. NFPA 1915, 2000 ed.

Steering Drive Ends. In the front wheel spindle in a driving–steering axle as used at the front of an all-wheel drive vehicle. NFPA 414, 2001 ed.

Stem. The most forward portion of the hull. NFPA 1925, 2004 ed.

Sterilization. The use of a physical or chemical procedure to destroy all microbial life, including highly resistant bacterial endospores. NFPA 1581, 2005 ed.

Stern. The after end of a boat or vessel. NFPA 1405, 2001 ed.

Stevedore. A person employed in the loading and unloading of ships, sometimes called a longshoreman. NFPA 1405, 2001 ed.

FIGURE S.12 Cutting Torch.

Storage Life. The date to remove from service a vapor-protective ensemble or individual element that has undergone proper care and maintenance in accordance with manufacturer's instructions but has not been used either in training or at actual incidents. NFPA 1991, 2005 ed.

Storage Occupancy. An occupancy used primarily for the storage or sheltering of goods, merchandise, products, vehicles, or animals. NFPA 5000, 2002 ed.

Storage Tank. *As applied to drycleaning plants:* A tank used for the storage of new or distilled solvent. NFPA 32, 2004 ed. *As applied to flammable and combustible liquids:* Any vessel having a liquid capacity that exceeds 227 L (60 gal), is intended for fixed installation, and is not used for processing. NFPA 30, 2003 ed.

Story. The portion of a building located between the upper surface of a floor and the upper surface of the floor or roof next above. NFPA 5000, 2002 ed.

Strainer. A device used in pump inlets or tank fill openings that prevents foreign materials that cannot pass through the pump without causing damage from entering the tank or pump. (See Figure S.13.) NFPA 1906, 2001 ed.

Strategy. *As applied to wildland fire fighting:* The general plan or direction selected to accomplish incident objectives. NFPA 1051, 2002 ed. *As applied to public fire and life safety:* A comprehensive organizational plan that is designed to eliminate or mitigate risks that endanger lives, health, property, or the environment through public fire and life safety education programs. NFPA 1035, 2005 ed.

FIGURE S.11 Staypoles.

Figure S.13 Strainer.

Street. A public thoroughfare that has been dedicated for vehicular use by the public and can be used for access by fire department vehicles. NFPA 101, 2003 ed.

Street Floor. A story or floor level accessible from the street or from outside a building at ground level, with the floor level at the main entrance located not more than three risers above or below ground level and arranged and utilized to qualify as the main floor. NFPA 101B, 2002 ed.

Stress Area. Those areas of the garment that are subjected to more wear including, but not limited to, crotches, knees, elbows, and shoulders. NFPA 1851, 2001 ed.

Stressed-Skin-Type Boom Section. A boom framework that is fabricated by the welding of metal into full box sections with internal torsional members. NFPA 1914, 2002 ed.

Strike Team. Specified combinations of the same kind and type of resources, with common communications and a leader. (SEE FIGURE S.14.) NFPA 1051, 2002 ed.

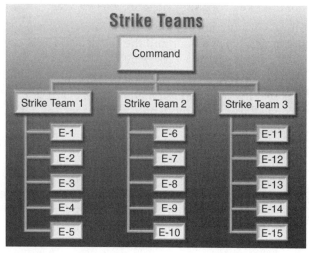

Figure S.14 Example of Strike Team Organization.

Strong-tight Packaging. Used for domestic shipment of materials with low levels of radioactivity with a low hazard and historical safety record such as depleted or natural uranium and rubble. NFPA 472, 2002 ed.

Strongback. The vertical members of a trench shoring system placed in contact with the earth, usually held in place against sections of sheeting with shores and positioned so that individual members do not contact each other. NFPA 1670, 2004 ed.

Structural Fire Fighting. The activities of rescue, fire suppression, and property conservation in buildings, enclosed structures, aircraft interiors, vehicles, vessels, aircraft, or like properties that are involved in a fire or emergency situation. NFPA 1710, 2004 ed.

Structural Fire Fighting Gloves. An element of the protective ensemble designed to provide minimum protection to the fingers, thumb, hand, and wrist. NFPA 1581, 2005 ed.

Structural Fire Fighting Protective Clothing. The protective clothing normally worn by fire fighters during structural fire fighting operations, which includes a helmet, coat, pants, boots, gloves, PASS device, and a hood to cover parts of the head not protected by the helmet and facepiece. NFPA 472, 2002 ed.

Structural Integrity. An unimpaired condition of any component. NFPA 1071, 2000 ed.

Structural Load Calculations. Load calculations based on the weight per cubic foot of construction materials such as concrete, steel, and wood. NFPA 1006, 2003 ed.

Structure Fire. Any fire inside, on, under, or touching a structure. NFPA 901, 2001 ed.

Structure Protection. To protect structures and other improvements from the threat of damage from an advancing wildland fire. NFPA 1051, 2002 ed.

Structure Protection Plan. A plan specifying safe and effective methods to protect structures and other improvements from the threat of damage from an advancing fire. NFPA 1051, 2002 ed.

Strut. The tensioned member placed between two opposing surfaces. NFPA 1006, 2003 ed.

Student. Any person who is present at the live fire training evolution for the purpose of receiving training. NFPA 1403, 2002 ed.

Subsidiary Communications Center (Subsidiary). A structure used to house a part of the control equipment of an emergency reporting system or communications system; also, a normally unattended facility that is remote from the communications center and is used to house equipment necessary for the functioning of an emergency communications system. NFPA 1221, 2002 ed.

Subterranean Rescue. Extraction from any environment natural or manmade that exists below grade as an enclosed environment with limited means of access or egress, including caves, tunnels, and mines. NFPA 1006, 2003 ed.

Suburban Area. As defined by the U.S. Census Bureau, an area with between 500 people and 1000 people per square mile. NFPA 1720, 2004 ed.

Suction Hose. A hose that is designed to prevent collapse under vacuum conditions so that it can be used for drafting water from below the pump (lakes, rivers, wells, etc.). NFPA 1961, 2002 ed.

Suction Lift. The sum of the vertical lift and the friction and entrance loss caused by the flow through the intake strainers and hose expressed in feet of water (meters of water) head. NFPA 1901, 2003 ed.

Suit Closure. The component that allows the wearer to enter (don) and exit (doff) the vapor-protective suit element. NFPA 1991, 2005 ed.

Suit Closure Assembly. The combination of the suit closure and the seam attaching the suit closure to the suit garment, excluding any protective flap or cover. NFPA 1991, 2005 ed.

Suit Material. The principal material used in the construction of the vapor-protective suit. NFPA 1991, 2005 ed.

Summarily Abate. To immediately judge a condition to be a fire hazard to life or property and to order immediate correction of such condition. NFPA 1, 2003 ed.

Sump. A recessed area of a tank assembly designed primarily to entrap sludge or debris for easy removal and to serve as a central liquid collection point. NFPA 1901, 2003 ed.

Superpressurization. The addition of gas to a fire extinguishing agent container to achieve a specified pressure therein. NFPA 2001, 2004 ed.

Superstructure. An enclosed structure above the main deck that extends from one side of the vessel to the other. NFPA 1405, 2001 ed.

Supervisor. An individual responsible for overseeing the performance or activity of other members. NFPA 1021, 2003 ed.

Supervisory Chief Officer. A member whose responsibility is to assume command through a formalized transfer of command process and to allow company officers to directly supervise personnel assigned to them. NFPA 1710, 2004 ed.

Supplemental Sheeting and Shoring. Sheeting and shoring operations that involve the use of commercial sheeting/shoring systems and/or isolation devices or that involve cutting and placement of sheeting and shoring when greater than two feet of shoring exists below the bottom of the strongback. NFPA 1670, 2004 ed.

Supplied Air Respirator (SAR). An atmosphere-supplying respirator for which the source of the breathing air is not designed to be carried by the user. Note: A supplied air respirator is also known as an "airline respirator." (See Figure S.15.) NFPA 1852, 2002 ed.

Supply Hose. Hose designed for the purpose of moving water between a pressurized water source and a pump that is supplying attack lines. NFPA 1961, 2002 ed.

Supply Line. One or more lengths of connected fire hose, also called a leader line, used to provide water to wyed lines or to the intake of a pump. NFPA 1410, 2005 ed.

Support System. A structure such as underpinning, bracing, or shoring that provides support to an adjacent structure, underground installation, or the sides of an excavation. NFPA 1670, 2004 ed.

Suppression. The sum of all the work done to extinguish a fire, beginning at the time of its discovery. NFPA 921, 2004 ed.

Surcharge Load. Any weight in the proximity of the trench that increases instability or the likelihood of secondary cave-in. NFPA 1006, 2003 ed.

Surface. A base that is secure and conducive to supporting and stabilizing a vehicle or object. NFPA 1006, 2003 ed.

FIGURE S.15 Supplied Air Respirator (SAR).

Surface Encumbrance. A natural or manmade structural object adjacent to or in the immediate vicinity of an excavation or trench. NFPA 1006, 2003 ed.

Surface Movement Guidance and Control System (SMGCS). A process or plan used by airports conducting operations in visibility conditions less than 366 m (1200 ft) runway visual range (RVR). NFPA 402, 2002 ed.

Surface Tension. The elastic-like force at the surface of a liquid, which tends to minimize the surface area, causing drops to form. NFPA 1150, 2004 ed.

Surface Water Rescue. Rescue of a victim who is afloat on the surface of a body of water. NFPA 1006, 2003 ed.

Surge. The elastic-like force at the surface of a liquid, which tends to minimize the surface area, causing drops to form. NFPA 1145, 2000 ed.

Surrogate Cylinder. A breathing gas cylinder only for testing in which the mass of the breathing air is replaced by a substitute mass. NFPA 1982, 1998 ed.

Suspension. The energy-attenuating system of the helmet made up of the headband and crown straps. NFPA 1976, 2000 ed.

Suspension System. The components utilized to attach the axle(s) to the frame assembly. NFPA 1912, 2001 ed.

Sustained Attack. The activities of fire confinement, control, and extinguishment that are beyond those assigned to the initial responding companies. NFPA 1710, 2004 ed.

Swash Partition. A vertical wall within a tank structure designed to control the unwanted movement of the fluid within that tank. NFPA 1901, 2003 ed.

Sweatband. That part of a helmet headband, either integral or attached, that comes in contact with the wearer's forehead. NFPA 1971, 2000 ed.

Swift Water. Water moving at a rate greater than 1 knot [1.85 km/hr (1.15 mph)]. NFPA 1670, 1999 ed.

Swim. To propel oneself through water by means of purposeful body movements and positioning. NFPA 1006, 2003 ed.

Swim Aids. Items of personal equipment that augment the individual rescuer's ability to propel through water. NFPA 1006, 2003 ed.

Switch. Any set of contacts that interrupts or controls current flow through an electrical circuit. NFPA 1901, 2003 ed.

Synthetic Breathing Air. A manufactured breathing air that is produced by blending nitrogen and oxygen. NFPA 1901, 2003 ed.

System. Several items of equipment assembled, grouped, or otherwise interconnected for the accomplishment of a purpose or function. NFPA 1, 2003 ed.

System Components. Life safety harness, belts, and auxiliary equipment devices. NFPA 1983, 2001 ed.

System Input. The input pressure or electrical power that the powered rescue tool is subjected to at any given moment. NFPA 1936, 2005 ed.

System Safety Check. A method of evaluating the safe assembly of a rescue system. NFPA 1670, 1999 ed.

System Safety Factor. The weakest point within a system, expressed as a ratio between the minimum breaking strength of that point (component) as compared to the force placed upon it. NFPA 1670, 2004 ed.

System Stress. Any condition creating excessive force (i.e., exceeding the maximum working load of any component) to components within a rope rescue system that could lead to damage or failure of the system. NFPA 1670, 2004 ed.

Tabletop Training. A workshop style of training involving a realistic emergency scenario and requiring problem-solving participation by personnel responsible for management and support at emergencies. NFPA 402, 2002 ed.

Tabulated Data. Any set of site-specific design data used by a professional engineer to design a protective system at a particular location. NFPA 1670, 2004 ed.

Tactical Considerations. Specific fire fighting objectives that are intended to support the strategy of the incident. NFPA 1710, 2004 ed.

Tactical Level Management Component (TLMC). A management unit identified in the incident management system commonly known as "division," "group," or "sector." NFPA 1561, 2005 ed.

Tactics. Deploying and directing resources on an incident to accomplish the objectives designated by strategy. NFPA 1051, 2002 ed.

Tactile Notification Appliance. A notification appliance that alerts by the sense of touch or vibration. NFPA 72, 2002 ed.

Tagout. A method of tagging, labeling, or otherwise marking an isolation device during hazard abatement operations to prevent accidental removal of the device. NFPA 1670, 2004 ed.

Tail Gasket. A gasket in the bowl of a coupling used to provide a watertight seal between the coupling and the hose in an expansion ring-type coupling. NFPA 1963, 1998 ed.

Talkgroup. A group of radios addressed as a single entity by the system and functionally equivalent to a conventional repeater channel. NFPA 1221, 2002 ed.

Tank Top. The lowest deck, top plate of the bottom tanks. NFPA 1405, 2001 ed.

Target Fuel. A fuel that is subject to ignition by thermal radiation such as from a flame or a hot gas layer. NFPA 921, 2004 ed.

Task. A specific job behavior or activity. NFPA 1002, 2003 ed.

Task Force. Any combination of single resources assembled for a particular tactical need, with common communications and a leader. NFPA 1051, 2002 ed.

Taxicab and Bus Repair Garages. Buildings, structures, or portions thereof used for storage, maintenance, and repair of fleets of taxicabs, sedan-limousine-type motor vehicles, or motor buses. *Note: Facilities for the dispensing of motor fuels are commonly provided in connection with these garages.* NFPA 1, 2000 ed.

Team. Two or more individuals who have been assigned a common task and are in communication with each other, coordinate their activities as a work group, and support the safety of one another. NFPA 1081, 2001 ed.

Technical Rescue. The application of special knowledge, skills, and equipment to safely resolve unique and/or complex rescue situations. NFPA 1670, 2004 ed.

Technical Rescue Incident. Complex rescue incidents requiring specially trained personnel and special equipment to complete the mission. NFPA 1670, 2004 ed.

Technical Specialist. Personnel with special skills that can be used anywhere within the IMS organization. NFPA 1561, 2005 ed.

Technician. An individual qualified and authorized by the compliant product manufacturer to provide specified care and maintenance to the product, and perform inspection, repair, and testing beyond the level classified as "user repair." NFPA 1852, 2002 ed.

Telecommunicator. An individual whose primary responsibility is to receive, process, or disseminate information of a public safety nature via telecommunication devices. (See Figure T.1.) NFPA 1061, 2002 ed.

Telescopic. Extended or retracted by sliding of the overlapping sections. NFPA 1914, 2002 ed.

Temperature. The degree of sensible heat of a body as measured by a thermometer or similar instrument. NFPA 921, 2004 ed.

Temporary Wiring. Approved wiring for power and lighting during a period of construction, remodeling, maintenance, repair, or demolition, and decorative lighting, carnival power and lighting, and similar purposes. NFPA 1, 2003 ed.

FIGURE T.1 Telecommunicator.

Tender. An individual trained in the responsibilities of diver safety who provides control of search patterns from the surface of the water. NFPA 1670, 2004 ed.

Terminal. As used in relation to computer-aided dispatching (CAD) systems/networks, an electronic device that combines a keyboard and a display screen to allow exchange of information between a telecommunicator and one or more computers in the system/network. NFPA 1221, 2002 ed.

Terminal. Either end of a carrier line having facilities for the handling of freight and passengers. NFPA 1405, 2001 ed.

Termination. That portion of incident management in which personnel are involved in documenting safety procedures, site operations, hazards faced, and lessons learned from the incident. *Note: Termination is divided into three phases: debriefing the incident, post-incident analysis, and critiquing the incident.* NFPA 1670, 2004 ed.

Terrain. Specific natural and topographical features within an environment. NFPA 1670, 2004 ed.

Terrain Hazard. Specific terrain feature, or feature-related condition, that exposes one to danger and the potential for injury and/or death. NFPA 1670, 2004 ed.

Test. To verify serviceability by measuring the mechanical, pneumatic, hydraulic, or electrical characteristics of an item and comparing those characteristics with prescribed standards. NFPA 1915, 2000 ed.

Test Lanyard. Static kernmantle rope used to connect the test mass to anchorage in dynamic drop tests. NFPA 1983, 2001 ed.

Tested. Verification of compliance with test requirements. NFPA 1932, 2004 ed.

Testing. The process by which the hazards that could confront entrants of a trench or excavation are identified and evaluated, including specifying tests that are to be performed in a trench or excavation. NFPA 1670, 2004 ed.

Textile Fabric. A planar structure consisting of yarns or fibers. NFPA 1971, 2000 ed.

Theoretical Critical Fire Area (TCA). A rectangle, the longitudinal dimension of which is the overall length of the aircraft, and the width includes the fuselage and extends beyond it by a predetermined set distance that is dependent on the overall width. Therefore, the aircraft length multiplied by the calculated width equals the size of the TCA. NFPA 402, 2002 ed.

Thermal Expansion. The proportional increase in length, volume, or superficial area of a body with rise in temperature. NFPA 921, 2004 ed.

Thermal Inertia. The properties of a material that characterize its rate of surface temperature rise when exposed to heat; related to the product of the material's thermal conductivity (k), its density (π), and its heat capacity (c). NFPA 921, 2004 ed.

Thermal Protective Clothing. Protective clothing such as helmets, footwear, gloves, hoods, trousers, and coats that are designed and manufactured to protect the fire brigade member from the adverse effects of fire. NFPA 1081, 2001 ed.

Thermal Protective Performance (TPP). A numerical value indicating the resistance of materials to convective and radiant heat exposure. NFPA 1977, 2005 ed.

Thermoplastic. Plastic materials that soften and melt under exposure to heat and can reach a flowable state. NFPA 921, 2004 ed.

Thermoset Plastics. Plastic materials that are hardened into a permanent shape in the manufacturing process and are not commonly subject to softening when heated; typically form char in a fire. NFPA 921, 2004 ed.

Thigh Circumference. Lower torso garment measurement at crotch line from folded edge to folded edge, and multiplied by 2 to obtain circumference. NFPA 1977, 2005 ed.

Thimble. A grooved metal reinforcement fitted snugly into an eye splice of rope to reduce chafing and to spread the tensional loads. NFPA 1925, 2004 ed.

Third Party Administrator (TPA). An organization contracted by a self-insured employer to handle the administrative aspects of the employer's plan. NFPA 1250, 2004 ed.

Thread Gasket. A gasket used in a female threaded connection to provide a watertight seal between the male and female threaded connections. NFPA 1963, 1998 ed.

Threshold. The beginning of that portion of the runway usable for landing. NFPA 402, 2002 ed.

Throat. The center of the footwear entrance area behind the gusset, from its top line to the lowest point where it attaches to the vamp. NFPA 1977, 2005 ed.

Throw Bag. A water rescue system that includes 15.24 m to 22.86 m (50 ft to 75 ft) of water rescue rope, an appropriately sized bag, and a closed-cell foam float. NFPA 1006, 2003 ed.

Throwline. A floating, one-person rope that is intended to be thrown to a person during water rescues or as a tether for rescuers entering the water. NFPA 1983, 2001 ed.

Thruster. Controllable devices used to assist in maneuvering and positioning the vessel. NFPA 1925, 2004 ed.

Tidal Water. Ocean water or bodies of water that are connected to oceans that either experience a twice daily rise and fall of their surface caused by the gravitational pull of the moon or experience a corresponding ebb and flow of water in response to the tides. *Note: Due to the connection to the ocean, all tidal water has some degree of salinity, which nontidal water lacks.* NFPA 1006, 2003 ed.

Tide Tables. Schedule of predicted rise and fall of the surface of tidal waters above or below a mean water level at predictable times of each day of the year. NFPA 1006, 2003 ed.

Tides. The periodic variation in the surface depth of the oceans, and of bays, gulfs, inlets, and tidal regions of rivers, caused by the gravitational pull of the sun and moon. NFPA 1405, 2001 ed.

Tie Circuit. A circuit that connects a communications center with an alternate communications center or with a public safety answering point (PSAP). NFPA 1221, 2002 ed.

Tiller Aerial Apparatus. A tractor-trailer aerial apparatus with a steering wheel connected to the rear axle for maneuvering the rear portion of the apparatus. NFPA 1002, 2003 ed.

Tiller Operator. The fire apparatus driver/operator. NFPA 1002, 2003 ed.

Time Line. Graphic representation of the events in a fire incident displayed in chronological order. NFPA 921, 2004 ed.

Tip. The end of the ladder opposite the butt end. (See Figure T.2.) NFPA 1931, 2004 ed.

Titanium. Refers to either pure metal or alloys having the generally recognized properties of titanium metal, including the fire or explosion characteristics of titanium in its various forms. NFPA 402, 2002 ed.

Figure T.2 Ladder Tip.

Toe. The point where the trench wall meets the floor of the trench. NFPA 1006, 2003 ed.

Toecap. A reinforcement to the toe area of footwear designed to protect the toes from impact and compression. NFPA 1971, 2000 ed.

Ton. Weight equivalent to 906 kg (2000 lb). NFPA 414, 2001 ed.

Tonnage. A measurement to determine vessel capacity (1 ton = 100 ft^3). NFPA 1925, 2004 ed.

Tool Kit. Equipment available to the rescuer. NFPA 1006, 2003 ed.

Top. The intersection between the midsagittal plane and the coronal plane extended to the helmet surface. NFPA 1971, 2000 ed.

Top Line. The top edge of the protective footwear, which includes the tongue, gusset, quarter, collar, and shaft. NFPA 1976, 2000 ed.

Top Rail. The top chord (rail) of an aerial ladder to which reinforcements are attached. NFPA 1901, 2003 ed.

Topographical Map. A graphical representation of the earth's surface, drawn to scale and reproduced in two dimensions, that reflects the topographical features of the area depicted. NFPA 1670, 1999 ed.

Topography. The land surface configuration. NFPA 1051, 2002 ed.

Torque Box. A structural component placed between the turntable and the chassis of an aerial device to absorb the stresses of operation. NFPA 1914, 2002 ed.

Torque Converter. A device that is similar to a fluid coupling but that produces, by means of additional turbine blades, variable torque multiplication. NFPA 414, 2001 ed.

Torque Value. A measure of tightness or the amount of stress that is put on a fastening device (i.e., bolt) to secure it properly. NFPA 1914, 2002 ed.

Total Continuous Electrical Load. The total current required to operate all of the devices permanently connected to the apparatus that can be simultaneously energized excluding intermittent-type loads such as primers and booster reel rewind motors. NFPA 1901, 2003 ed.

Total Flooding. The act and manner of discharging an agent for the purpose of achieving a specified minimum agent concentration throughout a hazard volume. NFPA 2001, 2004 ed.

Total Flooding System. A system consisting of an agent supply and distribution network designed to achieve a total flooding condition in a hazard volume. NFPA 2001, 2004 ed.

Total Quality Management (TQM). A management system fostering continuously improving performance at every level of function and focusing on customer satisfaction. NFPA 450, 2004 ed.

Towboat. A powerful, small vessel designed for pushing larger vessels. NFPA 1405, 2001 ed.

Toxic Gas. A gas with a median lethal concentration (LC50) in air of more than 200 ppm, but not more than 2000 ppm by volume of gas or vapor, or more than 2 mg/L, but not more than 20 mg/L of mist, fume, or dust, when administered by continuous inhalation for 1 hour (or less if death occurs within 1 hour) to albino rats weighing between 200 g and 300 g (0.44 lb and 0.66 lb) each. NFPA 55, 2005 ed.

Trace Number. A code that can be used to retrieve the production history of a product (e.g., a lot or serial number). NFPA 1999, 2003 ed.

Traditional Sheeting and Shoring. The use of 1.2 m × 2.4 m (4 ft × 8 ft) sheet panels, with a strongback attachment, supplemented by a variety of conventional shoring options such as hydraulic, screw, and/or pneumatic shores. NFPA 1670, 2004 ed.

Traffic Control. The direction or management of vehicle traffic such that scene safety is maintained and rescue operations can proceed without interruption. NFPA 1006, 2003 ed.

Traffic Control Devices. Ancillary equipment/resources used at the rescue scene to facilitate traffic control such as flares, barricades, traffic cones, or barrier tape. (See Figure T.3.) NFPA 1006, 2003 ed.

Training Center Burn Building. A structure specifically designed to conduct live fire training evolutions on a repetitive basis. NFPA 1403, 2002 ed.

Training Officer. The person designated by the fire chief with authority for overall management and control of the organization's training program. NFPA 1401, 2001 ed.

Transceiver. A combined transmitter and receiver unit. NFPA 1221, 2002 ed.

Transfer (9-1-1 call). The rerouting of a 9-1-1 call from one public safety answering point (PSAP) to another. NFPA 1221, 2002 ed.

Transfer Device. Various devices, including litters and harnesses, used with rope rescue systems to package and allow safe removal of a subject from a specific rescue environment. NFPA 1670, 2004 ed.

Transfer Pump. A separate engine or PTO-driven water pump mounted on the apparatus with a minimum rated capacity of 250 gpm (945 L/min) at 50 psi (345 kPa) net pump pressure and used primarily for water transfer. NFPA 1901, 1999 ed.

Travel Interval. The elapsed time starting when the responding vehicle wheels begin rolling toward the address or incident and ending when the vehicle arrives on scene at the address or incident location. NFPA 450, 2004 ed.

Treatment System. An assembly of equipment capable of processing a hazardous gas and reducing the gas concentration to a predetermined level at the point of discharge from the system to the atmosphere. NFPA 55, 2005 ed.

Figure T.3 Traffic Cones.

Trench Box (or Trench Shield). A manufactured protection system unit made from steel, fiberglass, or aluminum that is placed in a trench to protect workers from cave-in and that can be moved as a unit. NFPA 1670, 2004 ed.

Trench Emergency. Any failure of hazard control or monitoring equipment or other event(s) inside or outside a trench or excavation that could endanger entrants within the trench or excavation. NFPA 1670, 1999 ed.

Trench Floor. The bottom of the trench. NFPA 1006, 2003 ed.

Trench (or Trench Excavation). A narrow (in relation to its length) excavation made below the surface of the earth. NFPA 1670, 2004 ed.

Trench Upright. A vertical support member that spans the distance between the toe of the trench and the trench lip to collect and distribute the tension from the opposing wall over a large area. NFPA 1006, 2003 ed.

Trench/Excavation Functional Capability. The activity of removing a victim from a man-made cut, cavity, or depression in an earth surface, formed by earth removal. (See Figure T.4.) NFPA 1951, 2005 ed.

Triage. The sorting of casualties at an emergency according to the nature and severity of their injuries. NFPA 402, 2002 ed.

Triage Tag. A tag used in the classification of casualties according to the nature and severity of their injuries. (See Figure T.5.) NFPA 402, 2002 ed.

Trim. Retroreflective and fluorescent materials attached to the outermost surface of the protective ensemble element for visibility enhancement. *Note: Retroreflective materials enhance nighttime visibility, and fluorescent materials enhance daytime visibility.* NFPA 1976, 2000 ed.

Trouble Signal. A signal initiated by the fire alarm system or device indicative of a fault in a monitored circuit or component. NFPA 72, 2002 ed.

Trouser. An element of the protective ensemble that is designed to provide minimum protection to the lower torso and legs, excluding the ankles and feet. NFPA 1976, 2000 ed.

Truck. A common fire service term for aerial fire apparatus. NFPA 1410, 2005 ed.

Truck Company. A group of fire fighters who work as a unit and are equipped with one or more pieces of aerial fire apparatus. (See Figure T.6.) NFPA 1410, 2005 ed.

Trunk Line. A telephone line or channel between telephone central offices or switching devices, including lines between communications centers. NFPA 1221, 2002 ed.

Trunked Radio. A radio system that uses computer control to automatically assign channels from an available pool to users and groups of users. (See Figure T.7.) NFPA 1221, 2002 ed.

FIGURE T.4 Trench Rescue.

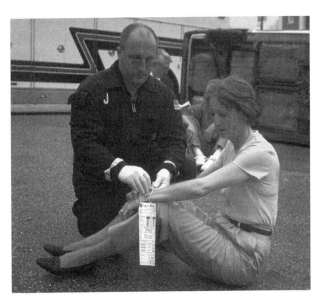

FIGURE T.5 Triage Tag.

FIGURE T.6 Truck Company Apparatus.

Trussed-Lattice-Type Boom Section. An open truss boom framework with vertical and diagonal braces that are fastened to horizontal beams of the frame. NFPA 1914, 2002 ed.

Turnaround. A portion of a roadway, unobstructed by parking, that allows for a safe reversal of direction for emergency equipment. NFPA 1144, 2002 ed.

Turning Clearance Radius. One-half the larger of the left or right full circle wall-to-wall turning diameter. NFPA 1901, 2003 ed.

Turnout Activation. Personnel preparation, boarding the vehicle, starting the vehicle, placing the vehicle in gear, and moving the vehicle toward the emergency scene. NFPA 450, 2004 ed.

Turnout Time (Turnout Interval). The time beginning when units acknowledge notification of the emergency to the beginning point of response time. NFPA 1710, 2004 ed.

Turnouts. A widening in a travelway of sufficient length and width to allow vehicles to pass one another. NFPA 1144, 2002 ed.

Turntable. A structural component that allows 360-degree continuous rotation of an aerial device through a rotating bearing and that connects the aerial device to the chassis and stabilization system and could contain an operator's control station. NFPA 1912, 2001 ed.

Turntable Alignment Indicator. An indicator that facilitates alignment of the aerial device with the boom support for bedding purposes. NFPA 1914, 2002 ed.

Turret. A vehicle-mounted master stream appliance. NFPA 402, 2002 ed.

Tween Decks. Cargo decks between main deck and lower hold. NFPA 1405, 2001 ed.

Twenty-Five Percent Drainage Time. See Foam Drain Time.

Twist. The degree of rotational movement from a given position. NFPA 1914, 2002 ed.

Two-Person Load. 272 kg (600 lb). NFPA 1006, 2003 ed.

Type 4 Rating. A rating for electrical equipment that is intended for outdoor use because it provides a degree of protection from falling rain, splashing water, and hose-directed water. NFPA 1901, 2003 ed.

Type A. Packaging for radioactive materials such as radiopharmaceuticals and low-level materials typically having an inner containment vessel of glass, plastic, or metal, and packaging materials made of polyethylene, rubber, or vermiculite. (See Figure T.8.) NFPA 472, 2002 ed.

FIGURE T.7 Trunked Radios.

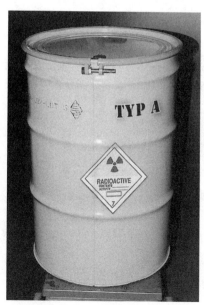

FIGURE T.8 Type A Package.

Type B. Packaging for radioactive materials such as spent fuel, high-level radioactive waste, and high concentrations of radioisotopes ranging from small drums (208 liter), Tru-Packs, to heavily shielded steel casks that can weigh more than 100 metric tons. (See Figure T.9.) NFPA 472, 2002 ed.

Figure T.9 Type B Package.

UAC. Abbreviation for the term "Universal Air Connection." Also known as RIC UAC. NFPA 1981, 2002 ed.

Ullage Hole. An opening in a tank hatch that allows measuring of liquid cargo. NFPA 1405, 2001 ed.

Ultimate Failure. Collapse of a ground ladder structure or component thereof. NFPA 1931, 2004 ed.

Ultimate Strength. The strength of a material in tension, compression, or shear, respectively, that is the maximum tensile, compressive, or shear stress that the material can sustain, calculated on the basis of the ultimate load and the original or unrestrained dimensions. NFPA 1901, 2003 ed.

Ultra High Frequency (UHF). Radio frequencies of 300 MHz to 3000 MHz. NFPA 1221, 2002 ed.

Ultrasonic Inspection. A nondestructive method of inspection in which high-frequency vibrations are injected through the surface of the test material and bounced back to their source from the opposite surface; if a flaw exists, signals return in a different pattern, revealing the location and extent of the flaw. NFPA 1914, 2002 ed.

UN/NA Identification Number. The four-digit number assigned to a hazardous material, which is used to identify and cross-reference products in the transportation mode. NFPA 472, 2002 ed.

Unauthorized Discharge. A release or emission of materials in a manner that does not conform to the provisions of NFPA 1 or applicable public health and safety regulations. NFPA 1, 2003 ed.

Underaxle Clearance. The clearance distance between the ground and the center drive train of the vehicle; generally this measurement is taken at the low point bottom of the drive differentials. NFPA 414, 2001 ed.

Underbody Clearance Dimensions. The dimensions determined with the vehicle fully loaded and fully equipped, unless otherwise specified. NFPA 414, 2001 ed.

Undercarriage. All components of an aircraft landing gear assembly. NFPA 402, 2002 ed.

Understanding. The process of gaining or developing the meaning of various types of materials or knowledge. NFPA 473, 2002 ed.

Undertruck Nozzle. A small nozzle device that hangs below the vehicle and disperses foam solution in a manner that provides protection for the vehicles from ground or grass proximity fires; these devices spray agent from wheel to wheel and front to back of the underside of the truck. NFPA 414, 2001 ed.

Unequipped Fire Apparatus. The completed fire apparatus excluding personnel, agent(s), and any equipment removable without the use of tools. NFPA 1901, 2003 ed.

Unified Command. A standard method to coordinate command of an incident where multiple agencies have jurisdiction. NFPA 1561, 2005 ed.

Uniform Corrosion. Removal of metal by chemical means over the entire surface. NFPA 1150, 2004 ed.

Uninterruptible Power Supply (UPS). A system designed to provide power, without delay or transients, during any period when the power source is incapable of performing. NFPA 1221, 2002 ed.

Unit. An engine company, truck company, or other functional or administrative group. NFPA 1021, 2003 ed.

Unit Operation or Unit Process. A segment of a physical or chemical process that might or might not be integrated with other segments to constitute the manufacturing sequence. NFPA 30, 2003 ed.

United Kingdom Civil Aviation Authority (CAA). An agency charged with the responsibility of regulating safety in civil aviation. NFPA 402, 2002 ed.

Unitized Rigid Body and Frame Structure. A structure in which parts that generally comprise a separate body are integrated with the chassis frame to form a single, rigid, load-carrying structure. NFPA 414, 2001 ed.

Universal Air Connection (UAC). The male fitting, affixed to the SCBA, and the female fitting, affixed to the filling hose, to provide emergency replenishment of breathing air to an SCBA breathing air cylinder. *Note: Universal air connection is also known as Rapid Intervention Crew/Company Universal Air Connection.* NFPA 1981, 2002 ed.

Universal Precaution. An approach to infection control in which human blood and certain human body fluids are treated as if known to be infectious for HIV, HBV, and other bloodborne pathogens. NFPA 1581, 2005 ed.

Unlined Hose. A hose consisting of only a woven jacket that is usually of linen yarns and is of such quality that the yarn swells when wet, tending to seal the hose. NFPA 1962, 2002 ed.

Unsprung Weight. The total weight of all vehicle components that are not supported completely by the suspension system. NFPA 414, 2001 ed.

Unstable Reactive Gas. A gas that, in the pure state or as commercially produced, will vigorously polymerize, decompose or condense, become self-reactive, or otherwise undergo a violent chemical change under conditions of shock, pressure, or temperature. NFPA 55, 2005 ed.

Unstable (Reactive) Material. A material that, in the pure state or as commercially produced, vigorously polymerizes, decomposes, or condenses, becomes self-reactive, or otherwise undergoes a violent chemical change under conditions of shock, pressure, or temperature. NFPA 1, 2003 ed.

Upgrade. The replacement or addition of components or systems with new components or systems with improved performance or capability. NFPA 1912, 2001 ed.

Upper. The part of footwear including, but not limited to, the toe, vamp, quarter, shaft, collar, and throat; but not including the sole with heel, puncture-resistant device, and insole. NFPA 1971, 2000 ed.

Upper Torso. The area of body above the waist and extending to the shoulders, including the arms and wrists but excluding the hands. NFPA 1971, 2000 ed.

Urban Area. As defined by the U.S. Census Bureau, an area with at least 1000 people per square mile. NFPA 1720, 2004 ed.

USAR Ensemble. The combination or assembly of multiple elements that are individually compliant with the USAR requirements of NFPA 1951 and that are designed to provide limited protection from the physical, environmental, thermal, chemical flash fire, chemical splash, and bloodborne hazards encountered during USAR operations. NFPA 1951, 2005 ed.

USAR Operations. Those technical incidents requiring at least one of the following: structural collapse functional capability, rope functional capability, confined space functional capability, trench/excavation functional capability, and vehicle/machinery functional capability, but not wilderness functional capability or water functional capability. NFPA 1951, 2005 ed.

USCG. United States Coast Guard. NFPA 1405, 2001 ed.

Use Condition I—Free Egress. Free movement is allowed from sleeping areas and other spaces where access or occupancy is permitted to the exterior via means of egress that meet the requirements of the Code. NFPA 1, 2003 ed.

Use Condition I—Free Egress—Detention and Correctional Occupancy. Free movement is allowed from sleeping areas and other spaces where access or occupancy is permitted to the exterior via means of egress that meet the necessary requirements. NFPA 101B, 2002 ed.

Use Condition II—Zoned Egress. Free movement is allowed from sleeping areas and any other occupied smoke compartment to one or more other smoke compartments. NFPA 1, 2003 ed.

Use Condition III—Zoned Impeded Egress. Free movement is allowed within individual smoke compartments, such as within a residential unit comprised of individual sleeping rooms and a group activity space, with egress impeded by remote-controlled release of means of egress from such a smoke compartment to another smoke compartment. NFPA 1, 2003 ed.

Use Condition IV—Impeded Egress. Free movement is restricted from an occupied space. Remote-controlled release is provided to allow movement from all sleeping rooms, activity spaces, and other occupied areas within the smoke compartment to another smoke compartment. NFPA 1, 2003 ed.

Use Condition V—Contained. Free movement is restricted from an occupied space. Staff-controlled manual release at each door is provided to allow movement from all sleeping rooms, activity spaces, and other occupied areas within the smoke compartment to another smoke compartment. NFPA 1, 2003 ed.

Use Condition—Contained—Detention and Correctional Occupancy. Free movement is restricted from an occupied space. Staff-controlled manual release at each door is provided to permit movement from all sleeping rooms, activity spaces, and other occupied areas within the smoke compartment to another smoke compartment. NFPA 101B, 2002 ed.

Utility Air. Air used for purposes other than human respiration. NFPA 1901, 2003 ed.

Utility Sink. A separate sink used for cleaning ensembles and ensemble elements. NFPA 1851, 2001 ed.

V

Vacant. No furnishings or equipment present. NFPA 901, 2001 ed.

Values at Risk. Public and private values that the wildland fire protection system is created and funded to protect. NFPA 1143, 2003 ed.

Valve Outlet Cap and Plug. A removable device that forms a gastight seal on the outlet to the control valve that is provided on a source containing a compressed gas or cryogenic fluid. NFPA 55, 2005 ed.

Valve Protection Cap. A rigid, removable cover provided for container valve protection during handling, transportation, and storage. NFPA 55, 2005 ed.

Valve Protection Device. A device attached to the neck ring or body of a cylinder for the purpose of protecting the cylinder valve from being struck or from being damaged by the impact resulting from a fall or an object striking the cylinder. NFPA 1, 2003 ed.

Vapor. The gas phase of a substance, particularly of those that are normally liquids or solids at ordinary temperatures. NFPA 921, 2004 ed.

Vapor Density. The ratio of the average molecular weight of a given volume of gas or vapor to the average molecular weight of an equal volume of air at the same temperature and pressure. (SEE FIGURE V.1.) NFPA 921, 2004 ed.

Vapor Pressure. The pressure, measured in pounds per square inch, absolute (psia), exerted by a liquid, as determined by ASTM D 323, Standard Method of Test for Vapor Pressure of Petroleum Products (Reid Method). (SEE FIGURE V.2.) NFPA 30, 2003 ed.

Vapor-Protective Clothing. The garment portion of a chemical-protective clothing ensemble that is designed and configured to protect the wearer against chemical vapors or gases. (SEE FIGURE V.3.) NFPA 472, 2002 ed.

Vapor-Protective Ensemble. Multiple elements of compliant protective clothing and equipment that when worn together provide protection from some risks, but not all risks, of vapor, liquid-splash, and particulate environments during hazardous materials incidents and from chemical and biological terrorism agents in vapor, gas, liquid, or particulate forms. NFPA 1991, 2005 ed.

Vapor-Protective Ensemble with Optional Chemical Flash Fire Escape and Liquefied Gas Protection. A compliant vapor-protective ensemble that is also certified as compliant with the optional requirements for both limited protection against chemical flash fire for escape only and for protection against liquefied gases. NFPA 1991, 2005 ed.

FIGURE V.1 Vapor Density.

FIGURE V.2 Vapor Pressure.

Figure V.3 Vapor-Protective Clothing.

Vapor-Protective Ensemble with Optional Chemical Flash Fire Escape Protection. A compliant vapor-protective ensemble that is also certified as compliant with the optional requirements for limited protection against chemical flash fire for escape only. NFPA 1991, 2005 ed.

Vapor-Protective Ensemble with Optional Liquefied Gas Protection. A compliant vapor-protective ensemble that is also certified as compliant with the optional requirements for protection against liquefied gases. NFPA 1991, 2005 ed.

Vapor-Protective Footwear. The ensemble element of the protective ensemble that provides chemical protection and physical protection to the feet, ankles, and lower legs. NFPA 1991, 2005 ed.

Vapor-Protective Gloves. The ensemble element of the protective ensemble that provides chemical protection to the hands and wrists. NFPA 1991, 2005 ed.

Vapor-Protective Suit. The ensemble garment element of the protective ensemble that provides chemical protection to the upper and lower torso, head, arms, and legs. NFPA 1991, 2005 ed.

Vehicle Safety Harness. A restraint device for vehicle occupants designed to limit their movement in the event of an accident, rapid acceleration, or rapid deceleration by securing individuals safely to a vehicle either in a seated position or tethered to the vehicle. NFPA 1500, 2002 ed.

Vehicle/Machinery Functional Capability. The activity of removing a victim from a vehicle or machine at an emergency incident. NFPA 1951, 2005 ed.

Vendor Confirmation. A written statement by the original manufacturer of a component that states the specification or performance range, or both, of the component. NFPA 1936, 2005 ed.

Vent. An opening for the passage of, or dissipation of, fluids, such as gases, fumes, smoke, and the like. NFPA 921, 2004 ed.

Ventilation. The changing of air within a compartment by natural or mechanical means. (See Figure V.4.) *Note: Ventilation can be achieved by introduction of fresh air to dilute contaminated air or by local exhaust of contaminated air.* NFPA 402, 2002 ed.

Ventilation-Controlled Fire. A fire in which the heat release rate or growth is controlled by the amount of air available to the fire. (See Figure V.5.) NFPA 921, 2004 ed.

Vertical Access Door. An access door installed in the vertical plane used to protect openings in fire-rated walls. NFPA 80, 1999 ed.

Vertical Circumference. One-piece garment measurement from junction of shoulder/collar seam down to the bottom of the crotch, and multiplied by 2 to obtain circumference. NFPA 1977, 2005 ed.

Vertical Stabilizer. That portion of the aircraft's empennage that contains the rudder. NFPA 402, 2002 ed.

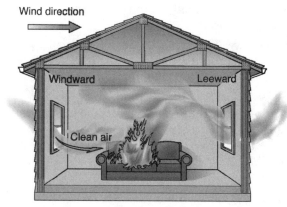

Figure V.4 Ventilation.

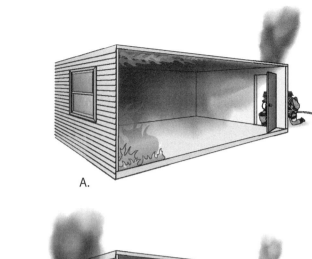
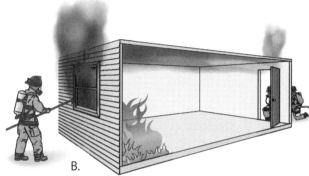

FIGURE V.5 A. Unvented Structure. B. Properly Vented Structure.

Vertical Zone. The area of a vessel between adjacent bulkheads. NFPA 1405, 2001 ed.

Very High Frequency (VHF). Radio frequencies of 30 MHz to 300 MHz. NFPA 1221, 2002 ed.

Vibration Isolation. Isolation materials used to prevent structure-borne vibrations from reaching attached surfaces. NFPA 1901, 2003 ed.

Victim Management. The manner of treatment given to those requiring rescue assistance. NFPA 1006, 2003 ed.

Victim Removal System. Those systems used to move a victim to a safe location. NFPA 1006, 2003 ed.

Virgin Fiber. Fiber that is new and previously unused. NFPA 1983, 2001 ed.

Viscosity. A measure of the resistance of a liquid to flow. NFPA 1150, 2004 ed.

Visible Damage. A permanent change in condition that is clearly evident by visual inspection without recourse to optical measuring or observation devices. NFPA 1931, 2004 ed.

Visor Material. The transparent chemical-protective material that allows the wearer to see outside the protective ensemble hood. NFPA 1991, 2005 ed.

Visual Inspection. Inspection by the eye without recourse to any optical devices, except prescription eyeglasses. NFPA 1914, 2002 ed.

Voice Communication Channel. A single path for transmitting electric signals that is distinct from other parallel paths. NFPA 1221, 2002 ed.

Volatile Liquid. A liquid that evaporates readily at normal temperature and pressure. NFPA 329, 2005 ed.

Volt (V). The unit of electrical pressure (electromotive force) represented by the symbol "E"; the difference in potential required to make a current of one ampere flow through a resistance of one ohm. NFPA 921, 2004 ed.

Volunteer Fire Department. A fire department having volunteer emergency service personnel comprising 85 percent or greater of its department membership. NFPA 1720, 2004 ed.

Waist. The area above the hips and below the xiphoid process. NFPA 1983, 2001 ed.

Waist Circumference. A garment measurement from top edge of waistband from folded edge to folded edge, and multiplied by 2 to obtain circumference. NFPA 1977, 2005 ed.

Wales. Horizontal members of a shoring system placed parallel to the excavation face whose sides bear against the vertical members of a shoring system or earth. *Note: Wales are also called walers or stringers.* NFPA 1006, 2003 ed.

Wall-to-Wall Turning Diameter. A measurement of the space that completely contains a vehicle as it is being turned. NFPA 414, 2001 ed.

Warm Zone. The control zone at a hazardous materials incident site where personnel and equipment decontamination and hot zone support take place. NFPA 472, 2002 ed.

Warp Thread. The threads or yarns of a hose reinforcement that run lengthwise to the hose. NFPA 1961, 2002 ed.

Waste Minimization. Treatment of hazardous spills by procedures or chemicals designed to reduce the hazardous nature of the material and to minimize the quantity of waste produced. NFPA 471, 2002 ed.

Water Hammer. The surge of pressure caused when a high-velocity flow of water is abruptly shut off. *Note: The pressure exerted by the flowing water against the closed system can be seven or more times that of the static pressure.* NFPA 1962, 2003 ed.

Water Hazard Zone. In water rescue, the zone includes the area covered by water or ice. NFPA 1670, 1999 ed.

Water Rescue Personal Protective Equipment. Personal equipment required to protect rescuers from physical dangers posed by exposure to in-water hazards and also those hazards that are associated with the climate and the adjacent area. NFPA 1006, 2003 ed.

Water Rescue Rope. Rope that floats, has adequate strength for anticipated use, is not weakened to the point of inadequacy for the task by saturation or immersion in water, and is of sufficient diameter to be gripped by bare wet hands. NFPA 1006, 2003 ed.

Water Supply. A source of water for fire fighting activities. NFPA 1144, 2002 ed.

Water Supply Officer (WSO). The fire department officer responsible for providing water for fire fighting purposes. NFPA 1142, 2001 ed.

Water Tower. An aerial device consisting of permanently mounted power-operated booms and a waterway designed to supply a large capacity mobile elevated water stream. *Note: The booms can be of articulating design or telescoping design.* NFPA 1901, 2003 ed.

Water-Bound Victim. A victim that is in the water needing assistance. NFPA 1006, 2003 ed.

Watercraft Conveyance. Devices intended for the purpose of transporting, moving, lifting, or lowering watercraft that may be required to be operated prior to and at the conclusion of every watercraft deployment. NFPA 1006, 2002 ed.

Watermanship Skills. Capabilities that include swimming, surface diving, treading water, and staying afloat with a reasonable degree of comfort appropriate to the required task. NFPA 1670, 2004 ed.

Watertight Bulkhead. A bulkhead (wall) strengthened and sealed to form a barrier against flooding in the event that the area on one side of it fills with liquid. NFPA 1405, 2001 ed.

Watertight Door. A door that is designed to keep water out. NFPA 1405, 2001 ed.

Watertight Transverse Bulkhead. A bulkhead through which there are no openings and that extends from the tank top up to the main deck, built to control flooding. NFPA 1405, 2001 ed.

Watt (W). Unit of power, or rate of work, equal to one joule per second, or the rate of work represented by a current of one ampere under the potential of one volt. NFPA 921, 2004 ed.

Wear Surface. The bottom of the footwear sole, including the heel. NFPA 1971, 2000 ed.

Weather Deck. Any deck that is exposed to the weather and normally accessible to personnel and that permits walking or moving around outboard of the superstructure. NFPA 1925, 2004 ed.

Weather Resistant. Sufficiently protected to prevent the penetration of rain, snow, and wind-driven sand, dirt, or dust under all operating conditions. NFPA 414, 2001 ed.

Webbing. Woven material of flat or tubular weave in the form of a long strip. NFPA 1670, 2004 ed.

Weft Thread. The threads or yarns of a hose reinforcement that are helically wound throughout the length of the hose at approximately right angles to the warp threads. NFPA 1961, 2002 ed.

Weight Scale Measurement. The accurate measurement of vehicle weight by means of a scale to verify or check a stated or estimated weight. NFPA 414, 2001 ed.

Weldment. A structure that is formed by the welding together of several components. NFPA 1914, 2002 ed.

Wet Chemical. An aqueous solution of organic or inorganic salts or a combination thereof that forms an extinguishing agent. NFPA 17A.

Wet Location. A nonsheltered location inside a compartment with a door or cover that, while open, exposes the electrical enclosure or panelboard to the same environmental conditions as the exterior of the fire apparatus. A location on a nonenclosed, exterior surface of a fire apparatus body or driving and crew compartment where the enclosure or panel is exposed to the environment. NFPA 1901, 2003 ed.

Wetting Ability. The ability of foam solution to penetrate and soak into a solid. NFPA 1150, 2004 ed.

Where Specified. Options selected by the purchaser beyond the minimum requirements of a standard. NFPA 414, 2001 ed.

Wilderness. An uncultivated, uninhabited, and natural area usually, but not necessarily, far from human civilization and trappings. NFPA 1670, 2004 ed.

Wildland. Land in an uncultivated, more or less natural state and covered by timber, woodland, brush, and/or grass. NFPA 901, 2001 ed.

Wildland Fire. An unplanned fire burning in vegetative fuels. (See Figure W.1.) NFPA 1051, 2002 ed.

Wildland Fire Apparatus. Fire apparatus designed for fighting wildland fires that is equipped with a pump having a capacity normally between 10 gpm and 500 gpm (38 L/min and 1900 L/min), a water tank, limited hose and equipment, and that has pump and roll capability. NFPA 1906, 2001 ed.

Figure W.1 Wildland Fire.

Wildland Fire Fighter I. The person at the first level of progression who has demonstrated the knowledge and skills necessary to function safely as a member of a wildland fire suppression crew under direct supervision. NFPA 1051, 2002 ed.

Wildland Fire Fighter II. The person at the second level of progression who has demonstrated the skills and depth of knowledge necessary to function under general supervision. NFPA 1051, 2002 ed.

Wildland Fire Fighting. The activities of fire suppression and property conservation in woodlands, forests, grasslands, brush, prairies, and other such vegetation, or any combination of vegetation, that is involved in a fire situation but is not within buildings or structures. NFPA 1500, 2002 ed.

Wildland Fire Fighting Chain Saw Protection. Items of protective equipment that provide protection to the legs, or to the lower torso and legs, excluding the ankles and feet. NFPA 1977, 2005 ed.

Wildland Fire Fighting Protective Clothing and Equipment. Items of compliant protective clothing and equipment products that provide protection from some risks, but not all risks, of emergency incident operations. NFPA 1977, 2005 ed.

Wildland Fire Fighting Protective Cold Weather Outerwear. Items of protective clothing that provide protection to the upper or lower torso, arms, and legs to provide insulation for warmth of the wearer during cold weather. NFPA 1977, 2005 ed.

Wildland Fire Fighting Protective Face/Neck Shroud. Item of protective clothing that provides protection to the face and neck area. (See Figure W.2.) NFPA 1977, 2005 ed.

Figure W.2 Wildland Fire Fighting Protective Face/Neck Shroud.

Wildland Fire Fighting Protective Footwear. Items of protective clothing that provides protection to the foot, ankle, and lower leg. NFPA 1977, 2005 ed.

Wildland Fire Fighting Protective Garments. Items of protective clothing that provide protection to the wearer's upper or lower torso, excluding the hands, face, and feet. NFPA 1977, 2005 ed.

Wildland Fire Fighting Protective Gloves. Item of protective clothing that provide protection to the hands and wrists. NFPA 1977, 2005 ed.

Wildland Fire Fighting Protective Goggles. Item of protective equipment that provides protection to the eyes and a portion of the face. NFPA 1977, 2005 ed.

Wildland Fire Fighting Protective Helmet. Item of protective equipment that provides protection to the head. NFPA 1977, 2005 ed.

Wildland Fire Fighting Protective Jacket. The protective outer garment item that provides protection to the upper torso and arms, excluding the hands and head. NFPA 1977, 2005 ed.

Wildland Fire Fighting Protective Load Carrying Equipment. Items of protective equipment that are worn by the wildland fire fighter to facilitate the carrying of gear. NFPA 1977, 2005 ed.

Wildland Fire Fighting Protective One-Piece Garment. The single-piece protective garment item that provides protection to the upper and lower torso, arms, and legs, excluding the head, hands, and feet. NFPA 1977, 2005 ed.

Wildland Fire Fighting Protective Shirt. A protective garment item that provides protection to the upper torso and arms, excluding the head and hands. NFPA 1977, 2005 ed.

Wildland Fire Fighting Protective Trousers. A protective garment item that provides protection to the lower torso and legs, excluding the feet. NFPA 1977, 2005 ed.

Wildland Fire Officer I. The person responsible for supervising and directing a single wildland fire suppression resource, such as a hand crew or an engine. NFPA 1051, 2002 ed.

Wildland Fire Officer II. The person responsible for commanding and managing resources in the suppression of all aspects of an extended attack wildland fire or an initial attack exceeding the capability of the Wildland Fire Officer I. NFPA 1051, 2002 ed.

Wildland/Urban Interface. Any area where wildland fuels threaten to ignite combustible homes and structures. NFPA 1143, 2003 ed.

Wildland/Urban Interface Coordinator. The person responsible for the development of the plan(s) for the reduction of the fire risks and hazards associated in the wildland/urban interface. NFPA 1501, 2002 ed.

Wildland/Urban Intermix. An area where improved property and wildland fuels meet with no clearly defined boundary. NFPA 1144, 2002 ed.

Winches. A stationary, motor-driven hoisting machine having a drum around which a rope or chain winds as the load is lifted. NFPA 1405, 2001 ed.

Windlass. A mechanical device utilized in the recovery of anchor and chain by vessels following anchoring operations. NFPA 1925, 2004 ed.

Window. Integral fabricated units, placed in an opening in a wall, primarily intended for the admission of light, or light and air, and not intended primarily for human entrance or exit. (See Figure W.3.) NFPA 80, 1999 ed.

Winter Liner. A garment term for an optional component layer designed to provide added insulation against cold. NFPA 1976, 2000 ed.

Winter Liner. An optional component layer for a garment designed to provide added insulation against cold. NFPA 1971, 2000 ed.

Wire Rope. Rope made of twisted strands of wire. NFPA 1670, 2004 ed.

Wired Circuit. A metallic circuit provided to or by a jurisdiction and dedicated to a specific alarm system that is under the control of or operated by, or is both under the control of and operated by, the jurisdiction or is shared with another jurisdiction. NFPA 1221, 2002 ed.

Wood Irregularities. Natural characteristics in or on the wood that can lower its durability, strength, or utility. NFPA 1931, 2004 ed.

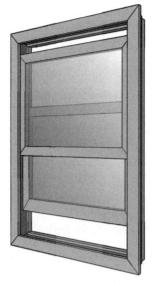  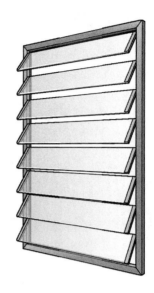

Figure W.3 Different Types of Windows.

Working Length. The length of a non-self-supporting portable ladder measured along the beams from the base support point of the ladder to the point of bearing at the top. NFPA 1932, 2004 ed.

Wristlet. *As applied to protective assemble for structural fire fighting:* An interface component element of the proximity protective ensemble that is the circular, close-fitting extension of the coat sleeve, usually made of knitted material, designed to provide limited protection to the protective coat-glove interface area. NFPA 1971, 2000 ed. *As applied to protective assemble for proximity fire fighting:* An interface component element of the protective ensemble that is the circular, close-fitting extension of the coat sleeve, usually made of knitted material, designed to provide limited protection to the protective coat/glove interface area. NFPA 1976, 2000 ed.

Written Notice. A notification in writing delivered in person to the individual or parties intended, or delivered at, or sent by certified or registered mail to, the last residential or business address of legal record. NFPA 1, 2003 ed.

Yield Strength. The stress at which a material exhibits a specified permanent deformation or set. NFPA 1901, 2003 ed.

APPENDIX A

Acronym Glossary

AAIB	Air Accident Investigations Branch
ACEP	American College of Emergency Physicians
ACLS	Advanced Cardiac Life Support
ACV	Air-Cushioned Vehicle
ADM	Alert Data Message
AED	Automated External Defibrillator
AHA	American Heart Association
AHJ	Authority Having Jurisdiction
ALI	Automatic Location Identification
ALOHA	Areal Locations of Hazardous Atmospheres
ALS	Advanced Life Support
ANI	Automatic Number Identification
ANSI	American National Standards Institute
ANSI/ASME	American National Standards Institute/American Society of Mechanical Engineers
APAS	Addressable Public Alerting System
APR	Air-Purifying Respirator
APU	Auxiliary Power Unit
ARC	American Red Cross
ARFF	Aircraft Rescue and Fire Fighting Vehicle
ASME	American Society of Mechanical Engineers
ASNT	American Society for Nondestructive Testing
ASTM	American Society of Testing and Materials
ATC	Airport Air Traffic Control
AVL	Automated Vehicle Locator
AWS	American Welding Society
BIA	Business Impact Analysis
BIPM	Bureau Internationale des Poids et Mesures
BLEVE	Boiling Liquid Expanding Vapor Explosion
BLS	Basic Life Support
Btu	British Thermal Unit
CA	Controlled Atmosphere
CAA	Civil Aviation Authority in United Kingdom
CAD	Computer-Aided Dispatch
CAFS	Compressed Air Foam System
CAMEO	Computer-Aided Management of Emergency Operations
CBRNE	Chemical, Biological, Radiological, Nuclear, or Explosive Material
CDR	Call Detail Recording
CEPPO	Chemical Emergency Preparedness and Prevention Office
CFR	Code of Federal Regulations
CGA	Compressed Gas Association
CISD	Critical Incident Stress Debriefing
CISM	Critical Incident Stress Management
CO	Carbon Monoxide
COP	Common Operational Picture
COTP	Coast Guard Captain of the Port
CPR	Cardiopulmonary Resuscitation
CQI	Continuous Quality Improvement
CRFFAA	Critical Rescue and Fire Fighting Access Area
CTO	Chief Technology Officer
CVR	Cockpit Voice Recorder
CW	Chemical Warfare
DEVS	Driver's Enhanced Vision System
DHS	Department of Homeland Security
DMAT	Disaster Medical Assistance Team
DOC	Department Operations Centers
DOT	Department of Transportation
DTRIM PCC	Domestic Threat Reduction and Incident Management Policy Coordination Center
ECFR	Emergency Care First Responder
EMAC	Emergency Management Assistance Compacts
EMD	Emergency Medical Dispatcher
EMS	Emergency Medical Service
EMS-C	Emergency Medical Services for Children
EMS/HM	Emergency Medical Services for Hazardous Materials
EMT	Emergency Medical Technician
EMT-A	Emergency Medical Technician-Ambulance
EMT-B	Emergency Medical Technician-Basic
EMT-I	Emergency Medical Technician-Intermediate
EMT-P	Emergency Medical Technician-Paramedic
ENS	Emergency Notification System
EOC	Emergency Operations Center
EOP	Emergency Operations Plan
ERA	Emergency Response Agency
ERF	Emergency Response Facility
ERP	Emergency Response Personnel
ESF	Emergency Support Functions
ESO	Emergency Services Organization

EVT	Emergency Vehicle Technician	LSF	Lowest Spreading Force
FAA	Federal Aviation Administration	M/A	Mechanical Advantage
FBI	Federal Bureau of Investigation	MACS	Multiagency Coordination System
FDR	Flight Data Recorder	MAP	Member Assistance Program
FEMA	Federal Emergency Management Agency	MEH	Mobile Emergency Hospital
FFFP	Film-Forming Fluoroprotein Foam	MIE	Minimum Ignition Energy
FLIR	Forward-Looking Infrared	MMRS	Metropolitan Medical Response System
FMVSS	Federal Motor Vehicle Safety Standard	MOU	Memoranda of Understanding
FOG	Field Operations Guide	MPE	Maximum Permissible Exposure
FP	Fluoroprotein Foam	MSDS	Materials Safety Data Sheet
FSS	Flight Service Station	MVA	Motor Vehicle Accident
FTC	Flight Technical Crew	NAEMSP	National Association of EMS Physicians
GAWR	Gross Axle Weight Rating	NAERG	North American Emergency Response Guidebook
GCWR	Gross Combustion Weight Rating	NCP	National Oil and Hazardous Substances Pollution Contingency Plan
GIS	Geographic Information System		
GPS	Global Positioning System	NDT	Nondestructive Testing
GSA	General Service Administration	NFPA	National Fire Protection Association
GVWR	Gross Vehicle Weight Rating	NGO	Nongovernmental Organization
Hc	Heat of Combustion	NH	National Hose Thread; National Standard Hose Thread
HIPAA	Health Insurance Portability and Accountability Act of 1996	NHTSA	National Highway Traffic Safety Administration
HMO	Health Maintenance Organization	NIC	NIMS Integration Center
HPF	Highest Pulling Force	NIH	National Institutes of Health
HRFP	Health-Related Fitness Program	NIMS	National Incident Management System
HRR	Heat Release Rate	NIOSH	National Institute for Occupational Safety and Health
HSF	Highest Spreading Force	NOAA	National Oceanic and Atmospheric Administration Office of Response and Restoration
HSI	Homeland Security Institute		
HSPD	Homeland Security Presidential Directive	NPAS	Nonaddressable Public Alerting System
HUD	Heads Up Display	NRP	National Response Plan
HVAC	Heating, Ventilation, and Air Conditioning Systems	NRP-CIA	National Response Plan Catastrophic Incident Annex
IAP	Incident Action Plan	NRP-CIS	National Response Plan Catastrophic Incident Supplement
IATA	International Air Transport Association	NTP	Normal Temperature and Pressure
IC	Incident Commander	NTSB	National Transportation and Safety Board
ICAO	International Civil Aviation Organization	NWCG	National Wildland Fire Coordinating Group
ICP	Incident Command Post	OEA	Oxygen-Enriched Atmosphere
ICS	Incident Command System	OIC	Office for Interoperability and Compatibility
IDLH	Immediately Dangerous to Life or Health	OPS	Operations Section Chief
IMS	Incident Management System	OPSEC	Operational Security
IMT	Incident Management Team	OSHA	Occupational Safety and Health Administration
INFOSEC	Information Security	PAS	Public Alerting System
IRIC	Initial Rapid Intervention Crew	PASS	Personal Alert Safety System
JAA	Joint Aviation Authority	PASS/SCBA	Personal Alert Safety System/Self-Contained Breathing Apparatus
JFO	Joint Field Office		
JIC	Joint Information Center	PCA	Practical Critical Fire Area
JIS	Joint Information System	PDD	Presidential Decision Directive
LC50	Lethal Concentration 50	PEL	Permissible Exposure Limit
LD50	Lethal Dosage 50	PFD	Personal Flotation Device
LEL	Lower Explosive Limit	PIO	Public Information Officer
LNG	Liquefied Natural Gas	PPE	Personal Protective Equipment
LO	Liaison Officer	PrPPE	Proximity Personal Protective Equipment
LP	Liquefied Petroleum	PSAP	Public Safety Answering Point
LPF	Lowest Pulling Force	PTO	Power Takeoff

PTT	Push-to-Talk	**SPCC**	Spill Prevention Control and Countermeasure Plan
QA	Quality Assessment	**STEL**	Short-Term Exposure Limit
R&D	Research and Development	**STP**	Standard Temperature and Pressure
RIC	Rapid Intervention Crew/Company	**SWOT**	Strengths, Weaknesses, Opportunities, and Threats Within a System
RIC UAC	Rapid Intervention Crew/Company Universal Air Connection System	**TC**	Transport Canada
ROC	Regional Operations Centers	**TCA**	Theoretical Critical Fire Area
RPE	Rate of Perceived Exertion; Respiratory Protective Equipment	**TLMC**	Tactical Level Management Component
RPP	Radiant Protective Performance	**TPA**	Third Party Administrator
SAR	Supplied Air Respirator	**TPP**	Thermal Protective Performance
SCBA	Self-Contained Breathing Apparatus	**TQM**	Total Quality Management
SCBA/SAR	Self-Contained Breathing Apparatus/Supplied Air Respirators	**UAC**	Universal Air Connection
scf	Standard Cubic Foot	**UC**	Unified Command
scfm	Standard Cubic Feet Per Minute	**UHF**	Ultra High Frequency
SCUBA	Self-Contained Underwater Breathing Apparatus	**UPS**	Uninterruptible Power Supply
SDO	Standards Development Organizations	**USAR**	Urban Search and Rescue Teams
SEMS	Standard Emergency Management System (California)	**USCG**	United States Coast Guard
SETIQ	Emergency Transportation System for the Chemical Industry in Mexico	**UTC**	Coordinated Universal Time
SMGCS	Surface Movement Guidance and Control System	**VHF**	Very High Frequency
SO	Safety Officer	**VLRA**	Valve-Regulated Lead-Acid Battery
SOLAS	International Convention for the Safety of Life at Sea	**WMD**	Weapons of Mass Destruction
SOP	Standard Operating Procedure	**WSO**	Water Supply Officer

APPENDIX B

Alphabetical Index—2006 NFC Contents

51A	Acetylene Cylinder Charging Plants—2001	Vol 3
30B	Aerosol Products, Manufacture and Storage—2002	Vol 3
422	Aircraft Accident/Incident Response Assessment—2004	Vol 14
407	Aircraft Fuel Servicing—2001	Vol 8
408	Aircraft Hand Portable Fire Extinguishers—2004	Vol 8
409	Aircraft Hangars—2004	Vol 8
410	Aircraft Maintenance—2004	Vol 8
402	Aircraft Rescue and Fire Fighting Operations—2002	Vol 14
403	Aircraft Rescue and Fire Fighting Services at Airports—2003	Vol 8
414	Aircraft Rescue and Fire-Fighting Vehicles—2001	Vol 8
1003	Airport Fire Fighter Professional Qualifications—2005	Vol 10
415	Airport Terminal Buildings, Fueling Ramp Drainage, Loading Walkways—2002	Vol 8
424	Airport/Community Emergency Planning—2002	Vol 14
101A	Alternative Approaches to Life Safety—2004	Vol 14
1901	Automotive Fire Apparatus—2003	Vol 12
85	Boiler and Combustion Systems Hazards—2004	Vol 5
1989	Breathing Air Quality for Fire and Emergency Services Respiratory Protection—2003	Vol 13
5000	Building Construction and Safety Code®—2006	Vol 13
900	Building Energy Code—2004	Vol 10
1402	Building Fire Service Training Centers—2002	Vol 15
12	Carbon Dioxide Extinguishing Systems—2005	Vol 1
720	Carbon Monoxide Warning Equipment in Dwellings—2005	Vol 9
1710	Career Fire Departments, Organization and Deployment—2004	Vol 11
211	Chimneys, Fireplaces, Vents, and Solid Fuel Burning Appliances—2003	Vol 7
260	Cigarette Ignition Resistance of Components of Upholstered Furniture—2003	Vol 7
1145	Class A Foams—2006	Vol 15
499	Classification of Combustible Dusts and of Hazardous (Classified) Locations for Electrical Installations in Chemical Process Areas—2004	Vol 14
497	Classification of Flammable Liquids, Gases, or Vapors and of Hazardous (Classified) Locations for Electrical Installations in Chemical Process Areas—2004	Vol 14
2001	Clean Agent Fire Extinguishing Systems—2004	Vol 13
1122	Code for Model Rocketry—2002	Vol 10
484	Combustible Metals—2006	Vol 9
1221	Communications, Emergency Services—2002	Vol 11
473	Competencies for EMS Personnel—2002	Vol 8
55	Compressed Gases and Cryogenic Fluids—2005	Vol 3
423	Construction and Protection of Aircraft Engine Test Facilities—2004	Vol 8
241	Construction, Alteration, and Demolition Operations—2004	Vol 7
306	Control of Gas Hazards on Vessels—2003	Vol 8
16	Deluge Foam-Water Sprinkler Systems and Foam-Water Spray Systems—2003	Vol 2
51	Design and Installation of Oxygen-Fuel Gas Systems for Welding, Cutting, and Allied Processes—2002	Vol 3
34	Dipping and Coating Processes Using Flammable or Combustible Liquids—2003	Vol 3
1600	Disaster/Emergency Management—2004	Vol 11
17	Dry Chemical Extinguishing Systems—2002	Vol 2
32	Drycleaning Plants—2004	Vol 3
850	Electric Generating Plants—2005	Vol 15
70B	Electrical Equipment Maintenance—2002	Vol 14
73	Electrical Inspection for Existing Dwellings—2006	Vol 5
70E	Electrical Safety in the Workplace—2004	Vol 5
79	Electrical Standard for Industrial Machinery—2002	Vol 5
731	Electronic Premises Security Systems—2006	Vol 9
110	Emergency and Standby Power Systems—2005	Vol 7
450	Emergency Medical Services and Systems—2004	Vol 14
1250	Emergency Service Organization Risk Management—2004	Vol 15
1561	Emergency Services Incident Management System—2005	Vol 11
1071	Emergency Vehicle Technicians—2006	Vol 10
560	Ethylene Oxide for Sterilization and Fumigation, Storage, Handling, and Use of—2002	Vol 9
412	Evaluating Aircraft Rescue and Fire-Fighting Foam Equipment—2003	Vol 8
555	Evaluating Potential for Room Flashover—2004	Vol 14
265	Evaluating Room Fire Growth Contribution of Textile Coverings on Full Height Panels and Walls—2002	Vol 7
551	Evaluation of Fire Risk Assessments—2004	Vol 14
285	Evaluation of Flame Propagation Characteristics of Exterior Non-Load-Bearing Wall Assemblies—2006	Vol 8
91	Exhaust Systems for Air Conveying of Gases, etc.—2004	Vol 6
69	Explosion Prevention Systems—2002	Vol 4
495	Explosive Materials Code—2006	Vol 9
80A	Exterior Fire Exposures—2001	Vol 14
268	Exterior Wall Assemblies, Test for Ignitability—2001	Vol 8
801	Facilities Handling Radioactive Materials—2003	Vol 9
705	Field Flame Test for Textiles and Films—2003	Vol 15
61	Fire and Dust Explosions in Agricultural and Food Processing Facilities—2002	Vol 4
921	Fire and Explosion Investigations, Guide for—2004	Vol 15
1002	Fire Apparatus Driver/Operator Professional Qualifications—2003	Vol 10
1915	Fire Apparatus Preventive Maintenance—2000	Vol 12
1912	Fire Apparatus Refurbishing—2001	Vol 12
1914	Fire Department Aerial Devices, Testing—2002	Vol 12
1581	Fire Department Infection Control Program—2005	Vol 11
1500	Fire Department Occupational Safety and Health Program—2002	Vol 11
13E	Fire Department Operations in Properties Protected by Sprinkler and Standpipe Systems—2005	Vol 14
1521	Fire Department Safety Officer—2002	Vol 11
80	Fire Doors and Fire Windows—1999	Vol 5
1001	Fire Fighter Professional Qualifications—2002	Vol 10
291	Fire Flow Testing and Marking of Hydrants—2002	Vol 14
1965	Fire Hose Appliances—2003	Vol 12
1963	Fire Hose Connections—2003	Vol 12
1961	Fire Hose—2002	Vol 12
1033	Fire Investigator Professional Qualifications—2003	Vol 10
1021	Fire Officer Professional Qualifications—2003	Vol 10
120	Fire Prevention and Control in Coal Mines—2004	Vol 7

Number	Title	Vol
122	Fire Prevention and Control in Metal/Nonmetal Mining and Metal Mineral Processing Facilities—2004	Vol 7
804	Fire Protection for Advanced Light Water Reactor Electric Generating Plants—2001	Vol 9
45	Fire Protection for Laboratories Using Chemicals—2004	Vol 3
914	Fire Protection in Historic Structures—2001	Vol 10
820	Fire Protection in Wastewater Treatment and Collection Facilities—2003	Vol 9
312	Fire Protection of Vessels During Construction, Conversion, Repair, and Lay-Up—2006	Vol 8
1401	Fire Protection Training Reports and Records—2001	Vol 15
703	Fire Retardant-Treated Wood and Fire-Retardant Coatings for Building Materials—2006	Vol 9
170	Fire Safety and Emergency Symbols—2006	Vol 7
550	Fire Safety Concepts Tree—2002	Vol 14
501A	Fire Safety Criteria for Manufactured Home Installations, Sites, and Communities—2005	Vol 9
150	Fire Safety in Racetrack Stables—2000	Vol 7
1041	Fire Service Instructor Professional Qualifications—2002	Vol 10
1983	Fire Service Life Safety Rope and System Components—2001	Vol 13
1000	Fire Service Professional Qualifications Accreditation and Certification System—2001	Vol 10
1404	Fire Service Respiratory Protection Training—2002	Vol 11
1451	Fire Service Vehicle Operations Training Program—2002	Vol 11
251	Fire Tests of Building Construction and Materials—2006	Vol 7
252	Fire Tests of Door Assemblies—2003	Vol 7
257	Fire Tests of Window and Glass Block Assemblies—2000	Vol 7
664	Fires and Explosions in Wood Processing and Woodworking Facilities—2002	Vol 9
1124	Fireworks and Pyrotechnic Articles—2006	Vol 11
1123	Fireworks Display—2006	Vol 10
2010	Fixed Aerosol Fire-Extinguishing Systems—2006	Vol 13
287	Flammability of Materials in Cleanrooms—2001	Vol 8
30	Flammable and Combustible Liquids Code—2003	Vol 3
288	Floor Fire Door Assemblies, Fire Tests of—2001	Vol 8
1150	Foam Chemicals for Fires in Class A Fuels—2004	Vol 11
97	Glossary of Terms Relating to Chimneys, Vents, and Heat Producing Appliances—2003	Vol 14
102	Grandstands, Folding and Telescopic Seating, Tents, and Membrane Structures—1995	Vol 7
12A	Halon 1301 Fire Extinguishing Systems—2004	Vol 1
329	Handling Releases of Flammable and Combustible Liquids and Gases—2005	Vol 14
99	Health Care Facilities—2005	Vol 6
1583	Health-Related Fitness for Fire Fighters—2000	Vol 11
271	Heat and Visible Smoke Release Rates for Materials and Products Using an Oxygen Consumption Calorimeter—2004	Vol 8
418	Heliports—2001	Vol 8
221	High Challenge Fire Walls, Fire Walls, and Fire Barrier Walls—2006	Vol 7
1127	High Power Rocketry—2002	Vol 11
851	Hydroelectric Generating Plants—2005	Vol 15
99B	Hypobaric Facilities—2005	Vol 6
704	Identification of the Hazards of Materials—2001	Vol 9
82	Incinerators, Waste and Linen Handling Systems and Equipment—2004	Vol 5
1081	Industrial Fire Brigade Member Professional Qualifications—2001	Vol 10
600	Industrial Fire Brigades—2005	Vol 9
2113	Industrial Personnel, Care of Flame-Resistant Garments—2001	Vol 13
2112	Industrial Personnel, Flame-Resistant Garments—2001	Vol 13
1962	Inspection, Care, Use of Fire Hose, Couplings, and Nozzles—2003	Vol 12
90A	Installation of Air-Conditioning and Ventilating Systems—2002	Vol 6
780	Installation of Lightning Protection Systems—2004	Vol 9
31	Installation of Oil-Burning Equipment—2001	Vol 3
13	Installation of Sprinkler Systems—2002	Vol 2
20	Installation of Stationary Pumps—2003	Vol 2
90B	Installation of Warm Air Heating and Air-Conditioning Systems—2006	Vol 6
1405	Land-Based Fire Fighters Who Respond to Marine Vessel Fires—2001	Vol 15
115	Laser Fire Protection—2003	Vol 7
101®	Life Safety Code®—2006	Vol 6
805	Light Water Reactor Electric Generating Plants—2001	Vol 9
59A	Liquefied Natural Gas (LNG)—2006	Vol 4
58	Liquefied Petroleum Gas Code—2004	Vol 3
430	Liquid and Solid Oxidizers—2004	Vol 8
1992	Liquid Splash-Protective Ensembles and Clothing for Hazardous Materials Emergencies—2005	Vol 13
1403	Live Fire Training Evolutions—2002	Vol 11
11	Low-, Medium-, and High- Expansion Foam—2005	Vol 1
35	Manufacture of Organic Coatings—2005	Vol 3
501	Manufactured Housing—2005	Vol 9
1931	Manufacturer's Design of Fire Department Ground Ladders—2004	Vol 12
303	Marinas and Boatyards—2006	Vol 8
1925	Marine Fire-Fighting Vessels—2004	Vol 12
307	Marine Terminals, Piers, and Wharves—2006	Vol 8
101B	Means of Egress—2002	Vol 7
270	Measurement of Smoke Obscuration Using a Conical Radiant Source—2002	Vol 8
1582	Medical Programs for Fire Departments—2003	Vol 11
301	Merchant Vessels—2001	Vol 8
261	Method of Test for Determining Resistance of Mock-Up Upholstered Furniture Material Assemblies to Ignition by Smoldering Cigarettes—2003	Vol 7
262	Method of Test for Flame Travel and Smoke of Wires and Cables—2002	Vol 7
272	Method of Test for Heat and Smoke Release Rates for Upholstered Furniture Components or Composites and Mattresses Using an Oxygen Consumption Calorimeter—2003	Vol 8
701	Methods of Fire Tests for Flame Propagation of Textiles and Films—2004	Vol 9
256	Methods of Fire Tests of Roof Coverings—2003	Vol 7
225	Model Manufactured Home Installation Standard—2005	Vol 7
1125	Model Rocket and High Power Rocket Motors—2001	Vol 11
140	Motion Picture and TV Production Facilities—2004	Vol 7
30A	Motor Fuel Dispensing Facilities—2003	Vol 3
610	Motorsports Venues—2003	Vol 15
70	National Electrical Code®—2005	Vol 4
72	National Fire Alarm Code®—2002	Vol 5
54	National Fuel Gas Code—2006	Vol 3
1852	Open-Circuit Self-Contained Breathing Apparatus (SCBA)—2002	Vol 11
1981	Open-Circuit Self-Contained Breathing Apparatus for Fire and Emergency Services—2002	Vol 12
1670	Operations and Training for Technical Rescue Incidents—2004	Vol 11
432	Organic Peroxide Formulations, Storage of—2002	Vol 8
86	Ovens and Furnaces—2003	Vol 5
53	Oxygen-Enriched Atmospheres—2004	Vol 14
88A	Parking Structures—2002	Vol 6
290	Passive Protection Materials for Use on LP-Gas Containers, Fire Testing—2003	Vol 8
1982	Personal Alert Safety Systems (PASS)—1998	Vol 13
274	Pipe Insulation Test to Evaluate Fire Performance Characteristics—2003	Vol 8
1141	Planned Building Groups—2003	Vol 11
302	Pleasure and Commercial Motor Craft—2004	Vol 8
10	Portable Fire Extinguishers—2002	Vol 1
505	Powered Industrial Trucks Including Type Designations, Areas of Use, Maintenance, and Operations—2006	Vol 9
1936	Powered Rescue Tools—2005	Vol 12
1620	Pre-Incident Planning—2003	Vol 15
730	Premises Security—2006	Vol 15
654	Prevention of Fire and Dust Explosions from Manufacturing Combustible Particulate Solids—2006	Vol 9
24	Private Fire Service Mains—2002	Vol 2
472	Professional Competence of Responders to Hazardous Materials Incidents—2002	Vol 8
1031	Professional Qualifications for Fire Inspector and Plan Examiner—2003	Vol 10
909	Protection of Cultural Resource Properties—2005	Vol 10
75	Protection of Information Technology Equipment—2003	Vol 5

No.	Title	Vol
1144	Protection of Life and Property from Wildfire—2002	Vol 11
318	Protection of Semiconductor Fabrication Facilities—2006	Vol 8
1977	Protective Clothing and Equipment for Wildland Fire Fighting—2005	Vol 12
1999	Protective Clothing for Medical Emergency Operations—2003	Vol 13
1976	Protective Ensemble for Proximity Fire Fighting—2000	Vol 12
1971	Protective Ensemble for Structural Fire Fighting—2000	Vol 12
1951	Protective Ensemble for USAR Operations—2001	Vol 12
1994	Protective Ensembles for Chemical/Biological Terrorism Incidents—2001	Vol 13
1201	Providing Emergency Services to the Public—2004	Vol 11
1035	Public Fire and Life Safety Educator Professional Qualifications—2005	Vol 10
1061	Public Safety Telecommunicator Qualifications—2002	Vol 10
496	Purged and Pressurized Enclosures for Electrical Equipment—2003	Vol 9
232	Records, Protection of—2000	Vol 7
1194	Recreational Vehicle Parks and Campgrounds—2005	Vol 11
1192	Recreational Vehicles—2005	Vol 11
405	Recurring Proficiency of Airport Fire Fighters—2004	Vol 8
1584	Rehabilitation of Members Operating at Incident Scene Operations and Training Exercises—2003	Vol 15
1006	Rescue Technician Professional Qualifications—2003	Vol 10
258	Research Test Method for Determining Smoke Generation of Solid Materials—2001	Vol 14
471	Responding to Hazardous Materials Incidents—2002	Vol 14
502	Road Tunnels, Bridges, and Other Limited Access Highways—2004	Vol 9
498	Safe Havens for Vehicles Transporting Explosives—2006	Vol 9
326	Safeguarding Tanks and Containers—2005	Vol 8
601	Security Services in Fire Loss Prevention—2005	Vol 9
1911	Service Tests of Fire Pump Systems on Fire Apparatus—2002	Vol 12
204	Smoke and Heat Venting—2002	Vol 7
105	Smoke Door Assemblies—2003	Vol 7
92B	Smoke Management Systems in Malls, Atria, Large Spaces—2005	Vol 6
92A	Smoke-Control Systems—2006	Vol 6
36	Solvent Extraction Plants—2004	Vol 3
33	Spray Application Using Flammable or Combustible Materials—2003	Vol 3
1964	Spray Nozzles—2003	Vol 12
13D	Sprinkler Systems in One- and Two-Family Dwellings and Manufactured Homes—2002	Vol 2
13R	Sprinkler Systems in Residential Occupancies up to and Including Four Stories in Height—2002	Vol 2
901	Standard Classifications for Incident Reporting and Fire Protection Data—2001	Vol 15
14	Standpipe and Hose Systems—2003	Vol 2
77	Static Electricity—2000	Vol 14
1975	Station/Work Uniforms—2004	Vol 12
37	Stationary Combustion Engines and Gas Turbines—2002	Vol 3
853	Stationary Fuel Cell Power Systems—2003	Vol 10
40	Storage and Handling of Cellulose Nitrate Film—2001	Vol 3
59	Storage and Handling of Liquefied Petroleum Gases at Utility Gas Plants—2004	Vol 4
490	Storage of Ammonium Nitrate—2002	Vol 9
434	Storage of Pesticides—2002	Vol 8
42	Storage of Pyroxylin Plastic—2002	Vol 3
111	Stored Electrical Energy Emergency and Standby Power Systems—2005	Vol 7
1851	Structural Fire Fighting Protective Ensembles—2001	Vol 11
520	Subterranean Spaces—2005	Vol 9
655	Sulfur Fires and Explosions—2001	Vol 9
385	Tank Vehicles for Flammable and Combustible Liquids—2000	Vol 8
76	Telecommunications Facilities—2005	Vol 5
253	Test for Critical Radiant Flux of Floor Covering Systems Using a Radiant Heat Energy Source—2006	Vol 7
259	Test Method for Potential Heat of Building Materials—2003	Vol 7
255	Test of Surface Burning Characteristics of Building Materials—2006	Vol 7
269	Toxic Potency Data, Test for Developing—2000	Vol 8
1452	Training Fire Department Personnel to Conduct Dwelling Fire Safety Surveys—2005	Vol 15
1410	Training for Initial Emergency Scene Operations—2005	Vol 11
130	Transit and Passenger Rail Systems—2003	Vol 7
220	Types of Building Construction—2006	Vol 7
1	Uniform Fire Code—2006	Vol 1
160	Use of Flame Effects Before an Audience—2006	Vol 7
1126	Use of Pyrotechnics Before a Proximate Audience—2006	Vol 11
1932	Use, Maintenance, and Service Testing of In-Service Fire Department Ground Ladders—2004	Vol 12
1991	Vapor-Protective Ensembles for Hazardous Materials Emergencies—2005	Vol 13
52	Vehicular Fuel Systems—2006	Vol 3
96	Ventilation Control and Fire Protection of Commercial Cooking Operations—2004	Vol 6
68	Venting of Deflagrations—2002	Vol 14
1720	Volunteer Fire Departments, Organization and Deployment—2004	Vol 11
286	Wall and Ceiling Interior Finish—2006	Vol 8
750	Water Mist Fire Protection Systems—2003	Vol 9
15	Water Spray Fixed Systems—2001	Vol 2
1142	Water Supplies for Suburban and Rural Fire Fighting—2001	Vol 11
22	Water Tanks for Private Fire Protection—2003	Vol 2
25	Water-Based Fire Protection Systems—2002	Vol 2
214	Water-Cooling Towers—2005	Vol 7
51B	Welding, Cutting, Other Hot Work—2003	Vol 3
17A	Wet Chemical Extinguishing Systems—2002	Vol 2
18	Wetting Agents—2006	Vol 2
1906	Wildland Fire Apparatus—2001	Vol 12
1051	Wildland Fire Fighter Professional Qualifications—2002	Vol 10
1143	Wildland Fire Management—2003	Vol 11

Numerical Index—2006 NFC Contents

1	Uniform Fire Code—2006	Vol 1
10	Portable Fire Extinguishers—2002	Vol 1
11	Low-, Medium-, and High-Expansion Foam—2005	Vol 1
12	Carbon Dioxide Extinguishing Systems—2005	Vol 1
12A	Halon 1301 Fire Extinguishing Systems—2004	Vol 1
13	Installation of Sprinkler Systems—2002	Vol 2
13D	Sprinkler Systems in One- and Two-Family Dwellings and Manufactured Homes—2002	Vol 2
13E	Fire Department Operations in Properties Protected by Sprinkler and Standpipe Systems—2005	Vol 14
13R	Sprinkler Systems in Residential Occupancies up to and Including Four Stories in Height—2002	Vol 2
14	Standpipe and Hose Systems—2003	Vol 2
15	Water Spray Fixed Systems—2001	Vol 2
16	Deluge Foam-Water Sprinkler Systems and Foam-Water Spray Systems—2003	Vol 2
17	Dry Chemical Extinguishing Systems—2002	Vol 2
17A	Wet Chemical Extinguishing Systems—2002	Vol 2
18	Wetting Agents—2006	Vol 2
20	Installation of Stationary Pumps—2003	Vol 2
22	Water Tanks for Private Fire Protection—2003	Vol 2
24	Private Fire Service Mains—2002	Vol 2
25	Water-Based Fire Protection Systems—2002	Vol 2
30	Flammable and Combustible Liquids Code—2003	Vol 3
30A	Motor Fuel Dispensing Facilities—2003	Vol 3
30B	Aerosol Products, Manufacture and Storage—2002	Vol 3
31	Installation of Oil-Burning Equipment—2001	Vol 3
32	Drycleaning Plants—2004	Vol 3
33	Spray Application Using Flammable or Combustible Materials—2003	Vol 3
34	Dipping and Coating Processes Using Flammable or Combustible Liquids—2003	Vol 3
35	Manufacture of Organic Coatings—2005	Vol 3
36	Solvent Extraction Plants—2004	Vol 3
37	Stationary Combustion Engines and Gas Turbines—2002	Vol 3
40	Storage and Handling of Cellulose Nitrate Film—2001	Vol 3
42	Storage of Pyroxylin Plastic—2002	Vol 3
45	Fire Protection for Laboratories Using Chemicals—2004	Vol 3
51	Design and Installation of Oxygen-Fuel Gas Systems for Welding, Cutting, and Allied Processes—2002	Vol 3
51A	Acetylene Cylinder Charging Plants—2001	Vol 3
51B	Welding, Cutting, Other Hot Work—2003	Vol 3
52	Vehicular Fuel Systems—2006	Vol 3
53	Oxygen-Enriched Atmospheres—2004	Vol 14
54	National Fuel Gas Code—2006	Vol 3
55	Compressed Gases and Cryogenic Fluids—2005	Vol 3
58	Liquefied Petroleum Gas Code—2004	Vol 3
59	Storage and Handling of Liquefied Petroleum Gases at Utility Gas Plants—2004	Vol 4
59A	Liquefied Natural Gas (LNG)—2006	Vol 4
61	Fire and Dust Explosions in Agricultural and Food Processing Facilities—2002	Vol 4
68	Venting of Deflagrations—2002	Vol 14
69	Explosion Prevention Systems—2002	Vol 4
70	National Electrical Code®—2005	Vol 4
70B	Electrical Equipment Maintenance—2002	Vol 14
70E	Electrical Safety in the Workplace—2004	Vol 5
72	National Fire Alarm Code®—2002	Vol 5
73	Electrical Inspection for Existing Dwellings—2006	Vol 5
75	Protection of Information Technology Equipment—2003	Vol 5
76	Telecommunications Facilities—2005	Vol 5
77	Static Electricity—2000	Vol 14
79	Electrical Standard for Industrial Machinery—2002	Vol 5
80	Fire Doors and Fire Windows—1999	Vol 5
80A	Exterior Fire Exposures—2001	Vol 14
82	Incinerators, Waste and Linen Handling Systems and Equipment—2004	Vol 5
85	Boiler and Combustion Systems Hazards—2004	Vol 5
86	Ovens and Furnaces—2003	Vol 5
88A	Parking Structures—2002	Vol 6
90A	Installation of Air-Conditioning and Ventilating Systems—2002	Vol 6
90B	Installation of Warm Air Heating and Air-Conditioning Systems—2006	Vol 6
91	Exhaust Systems for Air Conveying of Gases, etc.—2004	Vol 6
92A	Smoke-Control Systems—2006	Vol 6
92B	Smoke Management Systems in Malls, Atria, Large Spaces—2005	Vol 6
96	Ventilation Control and Fire Protection of Commercial Cooking Operations—2004	Vol 6
97	Glossary of Terms Relating to Chimneys, Vents, and Heat Producing Appliances—2003	Vol 14
99	Health Care Facilities—2005	Vol 6
99B	Hypobaric Facilities—2005	Vol 6
101®	Life Safety Code®—2006	Vol 6
101A	Alternative Approaches to Life Safety—2004	Vol 14
101B	Means of Egress—2002	Vol 7
102	Grandstands, Folding and Telescopic Seating, Tents, and Membrane Structures—1995	Vol 7
105	Smoke Door Assemblies—2003	Vol 7
110	Emergency and Standby Power Systems—2005	Vol 7
111	Stored Electrical Energy Emergency and Standby Power Systems—2005	Vol 7
115	Laser Fire Protection—2003	Vol 7
120	Fire Prevention and Control in Coal Mines—2004	Vol 7
122	Fire Prevention and Control in Metal/Nonmetal Mining and Metal Mineral Processing Facilities—2004	Vol 7
130	Transit and Passenger Rail Systems—2003	Vol 7
140	Motion Picture and TV Production Facilities—2004	Vol 7
150	Fire Safety in Racetrack Stables—2000	Vol 7
160	Use of Flame Effects Before an Audience—2006	Vol 7
170	Fire Safety and Emergency Symbols—2006	Vol 7
204	Smoke and Heat Venting—2002	Vol 7
211	Chimneys, Fireplaces, Vents, and Solid Fuel Burning Appliances—2003	Vol 7
214	Water-Cooling Towers—2005	Vol 7
220	Types of Building Construction—2006	Vol 7
221	High Challenge Fire Walls, Fire Walls, and Fire Barrier Walls—2006	Vol 7
225	Model Manufactured Home Installation Standard—2005	Vol 7
232	Records, Protection of—2000	Vol 7
241	Construction, Alteration, and Demolition Operations—2004	Vol 7
251	Fire Tests of Building Construction and Materials—2006	Vol 7
252	Fire Tests of Door Assemblies—2003	Vol 7
253	Test for Critical Radiant Flux of Floor Covering Systems Using a Radiant Heat Energy Source—2006	Vol 7
255	Test of Surface Burning Characteristics of Building Materials—2006	Vol 7
256	Methods of Fire Tests of Roof Coverings—2003	Vol 7
257	Fire Tests of Window and Glass Block Assemblies—2000	Vol 7
258	Research Test Method for Determining Smoke Generation of Solid Materials—2001	Vol 14
259	Test Method for Potential Heat of Building Materials—2003	Vol 7
260	Cigarette Ignition Resistance of Components of Upholstered Furniture—2003	Vol 7
261	Method of Test for Determining Resistance of Mock-Up Upholstered Furniture Material Assemblies to Ignition by Smoldering Cigarettes—2003	Vol 7
262	Method of Test for Flame Travel and Smoke of Wires and Cables—2002	Vol 7
265	Evaluating Room Fire Growth Contribution of Textile Coverings on Full Height Panels and Walls—2002	Vol 7
268	Exterior Wall Assemblies, Test for Ignitability—2001	Vol 8
269	Toxic Potency Data, Test for Developing—2000	Vol 8

No.	Title	Vol
270	Measurement of Smoke Obscuration Using a Conical Radiant Source—2002	Vol 8
271	Heat and Visible Smoke Release Rates for Materials and Products Using an Oxygen Consumption Calorimeter—2004	Vol 8
272	Method of Test for Heat and Smoke Release Rates for Upholstered Furniture Components or Composites and Mattresses Using an Oxygen Consumption Calorimeter—2003	Vol 8
274	Pipe Insulation Test to Evaluate Fire Performance Characteristics—2003	Vol 8
285	Evaluation of Flame Propagation Characteristics of Exterior Non-Load-Bearing Wall Assemblies—2006	Vol 8
286	Wall and Ceiling Interior Finish—2006	Vol 8
287	Flammability of Materials in Cleanrooms—2001	Vol 8
288	Floor Fire Door Assemblies, Fire Tests of—2001	Vol 8
290	Passive Protection Materials for Use on LP-Gas Containers, Fire Testing—2003	Vol 8
291	Fire Flow Testing and Marking of Hydrants—2002	Vol 14
301	Merchant Vessels—2001	Vol 8
302	Pleasure and Commercial Motor Craft—2004	Vol 8
303	Marinas and Boatyards—2006	Vol 8
306	Control of Gas Hazards on Vessels—2003	Vol 8
307	Marine Terminals, Piers, and Wharves—2006	Vol 8
312	Fire Protection of Vessels During Construction, Conversion, Repair, and Lay-Up—2006	Vol 8
318	Protection of Semiconductor Fabrication Facilities—2006	Vol 8
326	Safeguarding Tanks and Containers—2005	Vol 8
329	Handling Releases of Flammable and Combustible Liquids and Gases—2005	Vol 14
385	Tank Vehicles for Flammable and Combustible Liquids—2000	Vol 8
402	Aircraft Rescue and Fire Fighting Operations—2002	Vol 14
403	Aircraft Rescue and Fire Fighting Services at Airports—2003	Vol 8
405	Recurring Proficiency of Airport Fire Fighters—2004	Vol 8
407	Aircraft Fuel Servicing—2001	Vol 8
408	Aircraft Hand Portable Fire Extinguishers—2004	Vol 8
409	Aircraft Hangars—2004	Vol 8
410	Aircraft Maintenance—2004	Vol 8
412	Evaluating Aircraft Rescue and Fire-Fighting Foam Equipment—2003	Vol 8
414	Aircraft Rescue and Fire-Fighting Vehicles—2001	Vol 8
415	Airport Terminal Buildings, Fueling Ramp Drainage, Loading Walkways—2002	Vol 8
418	Heliports—2001	Vol 8
422	Aircraft Accident/Incident Response Assessment—2004	Vol 14
423	Construction and Protection of Aircraft Engine Test Facilities—2004	Vol 8
424	Airport/Community Emergency Planning—2002	Vol 14
430	Liquid and Solid Oxidizers—2004	Vol 8
432	Organic Peroxide Formulations, Storage of—2002	Vol 8
434	Storage of Pesticides—2002	Vol 8
450	Emergency Medical Services and Systems—2004	Vol 14
471	Responding to Hazardous Materials Incidents—2002	Vol 14
472	Professional Competence of Responders to Hazardous Materials Incidents—2002	Vol 8
473	Competencies for EMS Personnel—2002	Vol 8
484	Combustible Metals—2006	Vol 9
490	Storage of Ammonium Nitrate—2002	Vol 9
495	Explosive Materials Code—2006	Vol 9
496	Purged and Pressurized Enclosures for Electrical Equipment—2003	Vol 9
497	Classification of Flammable Liquids, Gases, or Vapors and of Hazardous (Classified) Locations for Electrical Installations in Chemical Process Areas—2004	Vol 14
498	Safe Havens for Vehicles Transporting Explosives—2006	Vol 9
499	Classification of Combustible Dusts and of Hazardous (Classified) Locations for Electrical Installations in Chemical Process Areas—2004	Vol 14
501	Manufactured Housing—2005	Vol 9
501A	Fire Safety Criteria for Manufactured Home Installations, Sites, and Communities—2005	Vol 9
502	Road Tunnels, Bridges, and Other Limited Access Highways—2004	Vol 9
505	Powered Industrial Trucks Including Type Designations, Areas of Use, Maintenance, and Operations—2006	Vol 9
520	Subterranean Spaces—2005	Vol 9
550	Fire Safety Concepts Tree—2002	Vol 14
551	Evaluation of Fire Risk Assessments—2004	Vol 14
555	Evaluating Potential for Room Flashover—2004	Vol 14
560	Ethylene Oxide for Sterilization and Fumigation, Storage, Handling, and Use of—2002	Vol 9
600	Industrial Fire Brigades—2005	Vol 9
601	Security Services in Fire Loss Prevention—2005	Vol 9
610	Motorsports Venues—2003	Vol 15
654	Prevention of Fire and Dust Explosions from Manufacturing Combustible Particulate Solids—2006	Vol 9
655	Sulfur Fires and Explosions—2001	Vol 9
664	Fires and Explosions in Wood Processing and Woodworking Facilities—2002	Vol 9
701	Methods of Fire Tests for Flame Propagation of Textiles and Films—2004	Vol 9
703	Fire Retardant-Treated Wood and Fire-Retardant Coatings for Building Materials—2006	Vol 9
704	Identification of the Hazards of Materials—2001	Vol 9
705	Field Flame Test for Textiles and Films—2003	Vol 15
720	Carbon Monoxide Warning Equipment in Dwellings—2005	Vol 9
730	Premises Security—2006	Vol 15
731	Electronic Premises Security Systems—2006	Vol 9
750	Water Mist Fire Protection Systems—2003	Vol 9
780	Installation of Lightning Protection Systems—2004	Vol 9
801	Facilities Handling Radioactive Materials—2003	Vol 9
804	Fire Protection for Advanced Light Water Reactor Electric Generating Plants—2001	Vol 9
805	Light Water Reactor Electric Generating Plants—2001	Vol 9
820	Fire Protection in Wastewater Treatment and Collection Facilities—2003	Vol 9
850	Electric Generating Plants—2005	Vol 15
851	Hydroelectric Generating Plants—2005	Vol 15
853	Stationary Fuel Cell Power Systems—2003	Vol 10
900	Building Energy Code—2004	Vol. 10
901	Standard Classifications for Incident Reporting and Fire Protection Data—2001	Vol 15
909	Protection of Cultural Resource Properties—2005	Vol 10
914	Fire Protection in Historic Structures—2001	Vol 10
921	Fire and Explosion Investigations, Guide for—2004	Vol 15
1000	Fire Service Professional Qualifications Accreditation and Certification System—2006	Vol 10
1001	Fire Fighter Professional Qualifications—2002	Vol 10
1002	Fire Apparatus Driver/Operator Professional Qualifications—2003	Vol 10
1003	Airport Fire Fighter Professional Qualifications—2005	Vol 10
1006	Rescue Technician Professional Qualifications—2003	Vol 10
1021	Fire Officer Professional Qualifications—2003	Vol 10
1031	Professional Qualifications for Fire Inspector and Plan Examiner—2003	Vol 10
1033	Fire Investigator Professional Qualifications—2003	Vol 10
1035	Public Fire and Life Safety Educator Professional Qualifications—2005	Vol 10
1041	Fire Service Instructor Professional Qualifications—2002	Vol 10
1051	Wildland Fire Fighter Professional Qualifications—2002	Vol 10
1061	Public Safety Telecommunicator Qualifications—2002	Vol 10
1071	Emergency Vehicle Technicians—2006	Vol 10
1081	Industrial Fire Brigade Member Professional Qualifications—2001	Vol 10
1122	Code for Model Rocketry—2002	Vol 10
1123	Fireworks Display—2006	Vol 10
1124	Fireworks and Pyrotechnic Articles—2006	Vol 11
1125	Model Rocket and High Power Rocket Motors—2001	Vol 11
1126	Use of Pyrotechnics Before a Proximate Audience—2006	Vol 11
1127	High Power Rocketry—2002	Vol 11
1141	Planned Building Groups—2003	Vol 11
1142	Water Supplies for Suburban and Rural Fire Fighting—2001	Vol 11
1143	Wildland Fire Management—2003	Vol 11
1144	Protection of Life and Property from Wildfire—2002	Vol 11
1145	Class A Foams—2006	Vol 15

No.	Title	Vol
1150	Foam Chemicals for Fires in Class A Fuels—2004	Vol 11
1192	Recreational Vehicles—2005	Vol 11
1194	Recreational Vehicle Parks and Campgrounds—2005	Vol 11
1201	Providing Emergency Services to the Public—2004	Vol 11
1221	Communications, Emergency Services—2002	Vol 11
1250	Emergency Service Organization Risk Management—2004	Vol 15
1401	Fire Protection Training Reports and Records—2001	Vol 15
1402	Building Fire Service Training Centers—2002	Vol 15
1403	Live Fire Training Evolutions—2002	Vol 11
1404	Fire Service Respiratory Protection Training—2002	Vol 11
1405	Land-Based Fire Fighters Who Respond to Marine Vessel Fires—2001	Vol 15
1410	Training for Initial Emergency Scene Operations—2005	Vol 11
1451	Fire Service Vehicle Operations Training Program—2002	Vol 11
1452	Training Fire Department Personnel to Conduct Dwelling Fire Safety Surveys—2005	Vol 15
1500	Fire Department Occupational Safety and Health Program—2002	Vol 11
1521	Fire Department Safety Officer—2002	Vol 11
1561	Emergency Services Incident Management System—2005	Vol 11
1581	Fire Department Infection Control Program—2005	Vol 11
1582	Medical Programs for Fire Departments—2003	Vol 11
1583	Health-Related Fitness for Fire Fighters—2000	Vol 11
1584	Rehabilitation of Members Operating at Incident Scene Operations and Training Exercises—2003	Vol 15
1600	Disaster/Emergency Management—2004	Vol 11
1620	Pre-Incident Planning—2003	Vol 15
1670	Operations and Training for Technical Rescue Incidents—2004	Vol 11
1710	Career Fire Departments, Organization and Deployment—2004	Vol 11
1720	Volunteer Fire Departments, Organization and Deployment—2004	Vol 11
1851	Structural Fire Fighting Protective Ensembles—2001	Vol 11
1852	Open-Circuit Self-Contained Breathing Apparatus (SCBA)—2002	Vol 11
1901	Automotive Fire Apparatus—2003	Vol 12
1906	Wildland Fire Apparatus—2001	Vol 12
1911	Service Tests of Fire Pump Systems on Fire Apparatus—2002	Vol 12
1912	Fire Apparatus Refurbishing—2001	Vol 12
1914	Fire Department Aerial Devices, Testing—2002	Vol 12
1915	Fire Apparatus Preventive Maintenance—2000	Vol 12
1925	Marine Fire-Fighting Vessels—2004	Vol 12
1931	Manufacturer's Design of Fire Department Ground Ladders—2004	Vol 12
1932	Use, Maintenance, and Service Testing of In-Service Fire Department Ground Ladders—2004	Vol 12
1936	Powered Rescue Tools—2005	Vol 12
1951	Protective Ensemble for USAR Operations—2001	Vol 12
1961	Fire Hose—2002	Vol 12
1962	Inspection, Care, Use of Fire Hose, Couplings, and Nozzles—2003	Vol 12
1963	Fire Hose Connections—2003	Vol 12
1964	Spray Nozzles—2003	Vol 12
1965	Fire Hose Appliances—2003	Vol 12
1971	Protective Ensemble for Structural Fire Fighting—2000	Vol 12
1975	Station/Work Uniforms—2004	Vol 12
1976	Protective Ensemble for Proximity Fire Fighting—2000	Vol 12
1977	Protective Clothing and Equipment for Wildland Fire Fighting—2005	Vol 12
1981	Open-Circuit Self-Contained Breathing Apparatus for Fire and Emergency Services—2002	Vol 12
1982	Personal Alert Safety Systems (PASS)—1998	Vol 13
1983	Fire Service Life Safety Rope and System Components—2001	Vol 13
1989	Breathing Air Quality for Fire and Emergency Services Respiratory Protection—2003	Vol 13
1991	Vapor-Protective Ensembles for Hazardous Materials Emergencies—2005	Vol 13
1992	Liquid Splash-Protective Ensembles and Clothing for Hazardous Materials Emergencies—2005	Vol 13
1994	Protective Ensembles for Chemical/Biological Terrorism Incidents—2001	Vol 13
1999	Protective Clothing for Medical Emergency Operations—2003	Vol 13
2001	Clean Agent Fire Extinguishing Systems—2004	Vol 13
2010	Fixed Aerosol Fire-Extinguishing Systems—2006	Vol 13
2112	Industrial Personnel, Flame-Resistant Garments—2001	Vol 13
2113	Industrial Personnel, Care of Flame-Resistant Garments—2001	Vol 13
5000	Building Construction and Safety Code®—2006	Vol 13

APPENDIX C

The National Incident Management System Glossary

Action messages Information that prompts the public to take immediate action.

Advisories and warnings Information that informs the public and provides specific instructions.

Air Operations Branch The component of the ICS responsible for all air resources. This includes both fixed- and rotor-wing aircraft, their support personnel, and landing areas.

Air Support Group Component of the Air Operations Branch that is responsible for all recordkeeping related to the aviation assets at an incident site.

Air Tactical Group Supervisor The individual responsible for the coordination of all airborne activity at an incident site.

Air-to-air nets Network for communications between aviation units.

ALOHA (Areal Locations of Hazardous Atmospheres) The chemical-plume-modeling component of the CAMEO software suite.

Area command A single command that manages multiple incident command posts. It is established for complex incidents that have span-of-control issues.

Area Command Logistics Chief Provides logistics support to the Area Commander and the related Incident Commander.

Area Command Planning Chief Provides planning support for the Area Commander and the related Incident Commanders.

Area Commander Responsible for the direction of incident management teams in a given area.

Assigned resource A resource that is engaged in supporting an incident.

Available resource A resource that is immediately capable of being assigned to a mission to support incident management operations.

Aviation Coordinator Coordinates aviation activities, including airspace management and resource prioritization, with the Area Commander.

Branches An area of incident management established to delegate an appropriate span of control under the Operations Section Chief.

CAMEO® (Computer Aided Management of Emergency Operations) A widely used system of software applications used to plan for and respond to chemical emergencies. It is one of the tools developed by EPA's Chemical Emergency Preparedness and Prevention Office (CEPPO) and the National Oceanic and Atmospheric Administration Office of Response and Restoration (NOAA) to assist frontline chemical emergency planners and responders.

Command net Network for communications among Command Staff, Section Chiefs, Branch Directors, and Division and Group Supervisors.

Command post The location from which the Incident Commander or Unified Command manage the incident. The command post should be easily identified and its location known to all responding resources.

Common operating picture (COP) The ability of incident management personnel in multiple locations to access common incident-related information. This information should be complete and include voice, data, and multimedia components.

Communications failure protocol Procedures for identifying major communication infrastructures and backup procedures in the case of system failures.

Communications Unit Plans the effective use of communications equipment and facilities assigned to an incident.

Compensation and Claims Unit A functional unit within the Finance/Administration Section that oversees and handles injury compensation and claims. This unit coordinates activities with the Medical Unit for on-scene care.

Corrective action plans Implemented solutions resulting in the reduction or elimination of an identified problem.

Cost Unit A functional unit within the Finance/Administration Section that is responsible for collecting, analyzing, and reporting all costs related to the management of an incident.

Demobilization Unit The unit in the Planning Section that develops the specific plan related to the release and return of resources to their original status.

Department of Homeland Security (DHS) The federal agency tasked with all aspects of domestic incident management.

Department Operations Center (DOC) An agency-specific center that coordinates with the EOC Operations Section.

Director of Emergency Management The senior manager of a local, county, or state emergency management agency.

Division A geographic area of an incident. Divisions are created to maintain span of control at large incident sites.

Documentation Unit The unit in the Planning Section that is responsible for all event documentation and administrative functions (copying, filing, etc.).

Domestic Reduction and Incident Management Policy Coordination Center (DTRIM PCC) A federal organization that has served as a national-level information hub for other incident management organizations.

Emergency Operations Center (EOC) A facility that serves as an incident support center and that displays a common operational picture of an incident or event.

Emergency Operations Plan (EOP) A systematic process to initiate, manage, and recover from any emergency in a similar manner to improve preparation and response.

Emergency responder As defined in the Homeland Security Act of 2002, Section 2(6), "The term 'emergency response providers' includes Federal, State, and local emergency public safety, law enforcement, emergency response, emergency medical (including hospital emergency facilities), and related personnel, agencies, and authorities." 6 U.S.C. 101(6).

Emergency Support Function (ESF) Annexes Portions of the NRP document that detail the functional processes and specialized applications based on the type of situation encountered. The Annexes outline the mission, policies, structure, and responsibilities of federal agencies in coordination with state or local governments.

Emergency Support Functions (ESFs) Components of an EOP that provide resources and protocols to support a particular aspect of incident management (e.g., evacuation or transportation).

Expendable resources Equipment, supplies, or tools that are normally used up or consumed in service or those that are more easily replaced than rescued, salvaged, or protected.

Facilities Unit The unit in the Operations Section that sets up, maintains, and demobilizes all facilities used in support of incident operations.

Finance/Administration Section The functional section of the Incident Command System responsible for financial reimbursement and administrative services to support an incident.

Finance/Administration Section Chief A member of the General Staff who provides cost estimates and arranges and ensures that the IAP is within the financial limits established by the Incident Commander.

Food Unit The unit in the Operations Section that plans food operations for facilities and sheltering operations.

Force protection Protection of key personnel and facilities to prevent losses in the event of an attack.

General Staff Consists of the following ICS positions: The Operations Section Chief, the Logistics Section Chief, the Planning Section Chief, the Financial/Administration Section Chief, and possibly an Intelligence Section Chief.

Ground Support Unit Maintains and repairs primary tactical equipment, vehicles, and mobile ground support equipment. Supplies fuel and provides transportation support.

Ground-to-air net Network for communications between ground and aviation units.

Groups A functional area of an incident. Groups are usually labeled according to their assigned job (e.g., Law Enforcement or Intelligence). Groups are not limited by geographic boundaries.

Horizontal coordination The coordination of various disciplines on the same level (e.g., local fire departments, police departments, and EMS agencies) so that they work together as a unit when responding to an incident.

ICS forms Forms developed by the National Wildfire Coordinating Group for ICS functions.

Incident Action Plan (IAP) A formal document that includes several components and provides a coherent means of communicating overall incident objectives in the contexts of both operational and support activities. The most important section is the incident objectives. An IAP is often verbal during fast-moving tactical events.

Incident Annexes Portions of the NRP that concentrate on specific hazardous situations that may require specialized application of the NRP.

Incident Command (IC) The ICS position responsible for overall incident management. This person establishes all strategic incident objectives and ensures that those objectives are carried out effectively.

Incident Command Post (ICP) A command center located at an incident site that focuses on tactical activities.

Incident Command System (ICS) A system for domestic incident management that is based on an expandable, flexible structure and that uses common terminology, positions, and incident facilities.

Incident Management Teams (IMTs) A group of incident management personnel carrying out the functions of Command, Operations, Planning, Logistics, Finance/Administration, and, in some cases, Intelligence.

Incident traffic plan A plan that specifies traffic routes and procedures for vehicles entering and departing an incident site, command post, or support area.

Information and intelligence function The component of NIMS that strives to provide the Incident Commander/Unified Command accurate and timely knowledge about a potential adversary and the surrounding operational environment.

Information sharing The development of a framework connecting various information systems, including incident notification and situation reports, status reporting, data analysis, geospatial information, wireless communications, and incident reports.

Intelligence Responsible for the collection and analysis of information related to the incident. Usually exists as part of the Command Staff, a section, a unit within the Planning Section, or as part of the Situation Unit, depending upon the scope and nature of the incident and the need for intelligence.

Interoperability The ability of separate entities to work together and coordinate effectively. In domestic incident management, interoperability refers to communications (radio) and data (computer) systems/programs. It also applies to equipment and technologies.

Joint Field Office (JFO) A temporary federal facility established to coordinate federal assistance to the affected jurisdiction during incidents of national significance.

Joint Information Center (JIC) The office within NIMS responsible for ensuring that all public information released about an incident is consistent and for screening any inappropriate facts that may damage an investigation.

Joint Information System (JIS) An integrated and coordinated mechanism to ensure the delivery of timely and accurate information.

Liaison Officer Maintains contact and coordination with off-incident support agencies related to the Area Command.

Liaison Officer (LNO) The Command Staff position responsible for providing a method of communication between the IC/UC and other supporting organizations.

Local government As defined in the Homeland Security Act of 2002, Section 2(10), the term "local government" means "(A) county, municipality, city, town, township, local public authority, school district, special district, intrastate district, council of governments (regardless of whether the council of governments is incorporated as a nonprofit corporation under State law), regional or interstate government entity, or agency or instrumentality of a local government; an Indian tribe or authorized tribal organization, or in Alaska a Native village or Alaska Regional Native Corporation; and a rural community, unincorporated town or village, or other public entity." 6 U.S.C. 101(10).

Logistics Section Responsible for all support requirements needed to facilitate effective incident management.

Management information systems Tools used to collect, update, and process data, track resources, and display their readiness status.

Marine nets Network for communications between marine units and land-based agencies.

Medical Unit Develops the incident medical plan and manages medical operations.

Milestone A benchmark with a specific deadline and measurable end-state.

Mitigation Reduction of harshness or hostility.

Mitigation plan A proposal to reduce or alleviate potentially harmful impacts. Any sustained action taken to reduce or eliminate the long-term risk to human life and property from hazards.

Multiagency Coordination System (MACS) A combination of facilities, equipment, personnel, procedures, and communications integrated into a common system for incident coordination and support that is usually located at an EOC and structured via the ICS template.

Mutual-aid agreement   An intergovernmental or interagency agreement that provides shared and common assistance when requested by member agencies. The equipment and personnel provided by a mutual-aid request may be predetermined for a particular type of incident or determined at the time of the request in consideration of available resources.

Narrative information   General information that informs the public about the nature and progress of an incident or event.

New normalcy   Concept developed by the Advisory Panel to Assess Domestic Response Capabilities for Terrorism Involving Weapons of Mass Destruction (commonly known as the Gilmore Commission). It calls upon America to develop a new, higher level of preparedness and attentiveness while still maintaining our commitment to the principles set forth by the Founding Fathers.

NIMS Integration Center (NIC)   The NIMS Integration Center (NIC) is a project of the Department of Homeland Security that is responsible for the ongoing development of NIMS, including defining those sections that are still primarily conceptual. Because the ongoing development of the NIMS is happening very quickly, organizations should routinely check for new information on the NIC Web site.

Nonexpendable resources   Equipment that is not normally used up or consumed in service or that can be easily recovered and made ready for continued service.

Office of Interoperability and Compatibility (OIC)   A DHS office chartered and launched on October 1, 2004. It is located within the Science and Technology Directorate of DHS. It develops interoperability standards and technologies for communications and data systems related to domestic incident management and assists local and state government organizations in achieving interoperability.

Operational security (OPSEC)   The protection of information that would compromise security or tactical operations.

Operations Section   Component of the ICS responsible for tactical operations at the incident site. The goal of the Operations Section is to reduce the immediate hazard, save lives and property, establish situational control, and restore normal conditions.

Operations Section Chief   The member of the General Staff who directly manages all incident tactical activities and implements the IAP. This individual usually has the greatest technical and tactical expertise with the incident problem.

Out-of-service resource   A resource that is unavailable.

Performance improvement and management programs   A program that provides quality management through assessment and corrective actions to remedy any deficiencies identified through the assessment process.

Physical resources   Personnel, teams, facilities, supplies, and major items of equipment available for assignment to or employment during incidents.

Planning Section   One of the major components of the ICS; this section collects, evaluates, and disseminates incident situation information and intelligence to the IC/UC and incident management personnel, prepares status reports, displays situation information, maintains status of resources assigned to the incident, and develops and documents the incident action plan based on guidance from the IC/UC.

Planning Section Chief   The supervisor of the Planning Section.

Policy Group   Elected officials and senior executives that give policy advice to the EOC Command.

Procurement Unit   A functional unit within the Finance/Administration Section that is responsible for the purchase of goods or services.

Public Information Officer (PIO)   Member of the ICS Command Staff responsible for gathering and releasing incident information to the media and other appropriate agencies.

Pull logistics   Ordering of personnel, supplies, and equipment from outside local response or support agencies.

Push logistics   Initial response equipment and supplies transported by responding units.

Quality management programs   A program that provides quality management through assessment and corrective actions to remedy any deficiencies identified through the assessment process.

Recovery plan   A guide for the activities to be undertaken by federal, state, or private entities to guide recovery efforts in areas affected by a disaster.

Regional operations centers (ROCs)   Agencies that coordinate the efforts of local EOCs. ROCs can support MACS.

Research and development (R&D)   The collection of information about a particular subject to create an action, process, tool, or result.

Resource management The coordination and oversight of assets that provide incident managers with timely and appropriate mechanisms to accomplish operational objectives during an incident.

Resource Unit The component of the Planning Section (in the ICS structure) that serves as the primary manager of all assigned personnel or other resources that have checked in at an incident site.

Resource Unit Leader The individual responsible for maintaining the status of all resources at an incident.

Resources Personnel, supplies, and equipment needed for incident operations.

Safety Officer (SO) The Command Staff position responsible for the management of the incident safety plan. This person has the authority to immediately stop any on-scene activity that is deemed to be unsafe.

Service Branch Provides communications, food, water, and medical services, and consists of the Communications Unit, Food Unit, and Medical Unit.

Single Command The Command structure in which a single individual is responsible for all of the strategic objectives of the incident. Typically used when an incident is within a single jurisdiction and is managed by a single discipline.

Single resource An individual response unit that is employed at or during an incident (e.g., a helicopter, water tanker, police vehicle, etc.).

Situation Unit This unit collects, processes, and organizes ongoing situation information; prepares situation summaries; and develops projections and forecasts of future events related to the incident. The Situation Unit also prepares maps and gathers and disseminates information and intelligence for use in the IAP.

Situation Unit Leader Monitors the status of objectives for each incident.

Situational awareness The ability to access all required information for effectively managing an incident. It is similar to the COP.

Staging area The location at which resources assigned to an incident are held until they are assigned to a specific function.

Standard operating procedures (SOPs) Detailed written procedures for the uniform performance of a function.

Standards development organizations (SDOs) Organizations with long-standing interest or expertise on existing approaches to standards for equipment and systems. Examples include the National Institute for Standards and Technology (NIST), the National Institute for Occupational Safety and Health (NIOSH), the American National Standards Institute (ANSI), and the National Fire Protection Association (NFPA).

State As defined in the Homeland Security Act of 2002, the term "State" means any State of the United States, the District of Columbia, the Commonwealth of Puerto Rico, the Virgin Islands, Guam, American Samoa, the Commonwealth of the Northern Mariana Islands, and any possession of the United States. 6 U.S.C. 101(14).

Strike Team A set number of the same type of resource operating under one leader.

Supply Unit Orders, receives, and processes all incident-related resources, personnel, and supplies.

Support Annexes Portions of the NRP that provide guidance to ensure that functions are carried out efficiently and effectively.

Support branch Provides services that assist incident operations by providing supplies, facilities, transport, and equipment maintenance, and consists of the Supply Unit, Facilities Unit, and Ground Support Unit.

Support net Communications network that supports logistics requests, resource status changes, and other nontactical functions.

SWOT An analysis tool based on examining strengths, weaknesses, opportunities, and threats.

Tactical net Communication network that connects operating agencies and functional units.

Task Force A combination of resources that work together to complete a specific mission.

Technical Specialists Personnel activated on an as needed basis who bring specific skills or knowledge to the incident management effort.

Technical Unit A unit within the Planning Section that can be set up to house Technical Specialists who will have long-term commitments to the incident and who will provide ongoing information to the incident management effort.

Time Unit A functional unit within the Finance/Administration Section that is responsible for ensuring the proper daily recording of personnel time. This unit also may track equipment usage time.

Total Quality Management (TQM) A program that provides quality management through assessment and corrective actions to remedy any deficiencies identified through the assessment process.

Treatment area An incident facility that is used as a location for the collection and treatment of patients prior to transport. Typically organized according to patient status.

Undersecretary for Science and Technology The position within the DHS responsible for working with the NIMS Integration Center to coordinate the establishment of technical and technology standards for NIMS users.

Unified area command An area command that spans multiple jurisdictions and gives each jurisdiction appropriate representation.

Unified Command (UC) The Command structure in which multiple individuals are cooperatively responsible for all the strategic objectives of the incident. Typically used when an incident is within multiple jurisdictions and/or is managed by multiple disciplines.

Unified logistics Utilization and coordination of two or more agencies or jurisdictions to manage diverse logistics functions.

Unified Operations Two or more Operations Section Chiefs from different disciplines or jurisdictions assigned to manage incident operations as a coordinated team.

Unified Planning The utilization of two or more planners from different agencies, disciplines, or jurisdictions functioning as a coordinated planning section (similar to the UC concept).

Unit A component of the ICS that is subordinate to a section.

Vertical coordination The coordination of different levels of government (local, state, federal, tribal) working together in incident management.

Fire Organization

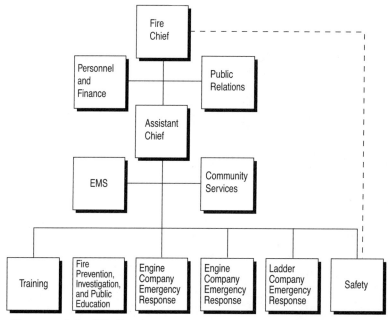

FIGURE D.1 Typical Organizational Structure of a Small Fire Department.

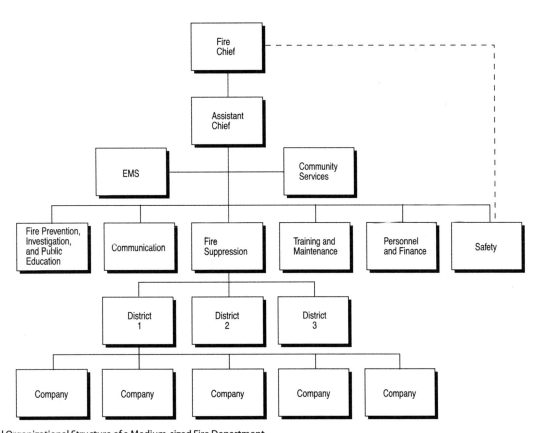

FIGURE D.2 Typical Organizational Structure of a Medium-sized Fire Department.

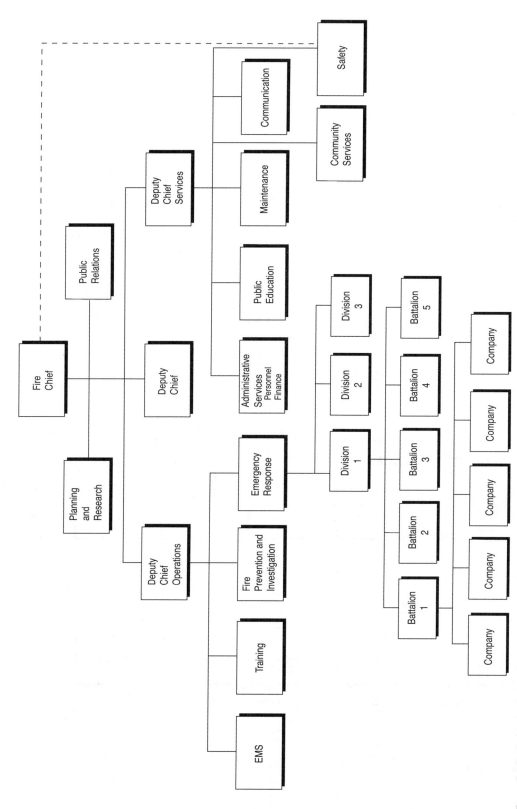

FIGURE D.3 Typical Organizational Structure of a Large Fire Department.

APPENDIX E

National Fire Associations and Organizations

Air and Surface Transport Nurses Association
9101 East Kenyon Avenue
Suite 3000
Denver, CO 80237
800-897-6362

Aircraft Rescue and Firefighting Working Group
1701 West Northwest Highway
Grapevine, TX 76051
817-329-5092

American Ambulance Association
8201 Greensboro Drive
Suite 300
McLean, VA 22102

American Association for Respiratory Care
9425 North MacArthur Blvd
Suite 100
Irving, TX 75063
972-243-2272

American Association of Critical Care Nurses
101 Columbia
Aliso Viejo, CA 92656
949-362-2000

American Board of Emergency Medicine
3000 Coolidge Road
East Lansing, MI 48823
517-332-4800

American Burn Association
625 North Michigan Avenue
Suite 1530
Chicago, IL 60611
312-642-9260

American College of Emergency Physicians
1125 Executive Circle
Irving, TX 75038
800-798-1822

American College of Surgeons
633 North Saint Clair Street
Chicago, IL 60611
312-202-5000

American Osteopathic Board of Emergency Medicine
142 East Ontario Street
Fourth Floor
Chicago, IL 60611
312-335-1065

American Red Cross
2025 E Street, NW
Washington, DC 20006
202-303-4498

American Trauma Society
8903 Presidential Parkway
Suite 512
Upper Marlboro, MD 20772
800-556-7890

Association of Air Medical Services
526 King Street
Suite 415
Alexandria, VA 22314
703-836-8732

Association of Public-Safety Communications Officials
351 North Williamson Blvd
Daytona Beach, FL 32114
386-322-2500

ASTM International
100 Barr Harbor Drive
PO Box C700
West Conshohocten, PA 19428
610-832-9500

Brain Injury Association of America
8201 Greensboro Drive
Suite 611
McLean, VA 22102
703-761-0750

Central Station Alarm Association
440 Maple Avenue East
Suite 201
Vienna, VA 22180
703-242-4670

Commission on Accreditation of Ambulance Services
1926 Waukegan Road
Suite 1
Glenview, IL 60025
847-657-6828

Commission on Accreditation of Medical Transport Systems (CAMTS)
PO Box 130
Sandy Springs, SC 29677
864-287-4177

Congressional Fire Service Institute
900 Second Street, NE
Suite 303
Washington, DC 20002
202-371-1277

Doctors for Disaster Preparedness
1601 North Tucson Blvd
Suite 9
Tucson, AZ 85716
520-325-2680

ECRI
5200 Butler Pike
Plymouth Meeting, PA 19462
610-825-6000

Emergency Management Institute
U.S. Fire Administration
16825 S. Seton Avenue
Emmitsburg, MD 21727
301-447-1251

Emergency Medical Service Institute
221-2500 Penn Avenue
Pittsburgh, PA 15221
412-242-7322

Emergency Medicine Residents Association
1125 Executive Circle
Irving, TX 75038
972-550-0920

Emergency Nurses Association (ENA)
915 Lee Street
Des Plaines, IL 60016
800-900-9659

Federal Emergency Management Agency (FEMA)
500 C Street, SW
Washington, DC 20472
202-566-1600

Fire Apparatus Manufacturers Association (FAMA)
PO Box 397
Lynnfield, MA 01940
781-334-2911

Fire Department Safety Officers Association
PO Box 149
Ashland, MA 01721
508-881-3114

Helicopter Association International
1635 Prince Street
Alexandria, VA 22314
703-683-4646

International Association of Arson Investigators (IAAI)
12770 Boenker Road
Bridgeton, MO 63044
314-739-4224

International Association of Black Professional Firefighters
1020 North Taylor Avenue
St. Louis, MO 63113
786-229-6914

International Association of Emergency Managers
201 Park Washington Court
Falls Church, VA 22046
703-538-1795

International Association of Fire Chiefs (IAFC)
4025 Fair Ridge Drive
Fairfax, VA 22033
703-273-0911

International Association of Firefighters (IAFF)
1750 New York Avenue, NW
Washington, DC 20006
202-737-8484

International Association of Flight Paramedics
4835 Riveredge Cove
Snellville, GA 30039
770-979-6372

International Code Council
5203 Leesburg Pike, Suite 600
Falls Church, VA 22041
888-422-7233

International Critical Incident Stress Foundation Inc
3290 Pine Orchard Lane
Suite 106
Ellicott City, MD 21042
410-750-9600

International Fire Marshals Association (IFMA)
National Fire Protection Association
One Batterymarch Park
Quincy, MA 02169
617-770-3000

International Rescue and Emergency Care Association (IRECA)
PO Box 431000
Minneapolis, MN 55443
800-854-7322

International Trauma Life Support
1 S. 280 Summit Avenue, Court B-2
Oakbrook Terrace, IL 60182
800-495-6442

Mountain Rescue Association
PO Box 880868
San Diego, CA 92168
619-884-9456

National Association for Search and Rescue (NASAR)
PO Box 232020
Centreville, VA 20120
703-222-6277

National Association of Emergency Medical Technicians (NAEMT)
PO Box 1400
Clinton, MS 39056
800-34-NAEMT

National Association of EMS Educators (NAEMSE)
681 Andersen Drive
Foster Plaza 6
Pittsburgh, PA 15220
412-920-4775

National Association of EMS Physicians
PO Box 15945-281
Lenexa, KS 66285
913-492-5858

National Association of Fire Investigators
857 Tallevast Road
Sarasota, FL 34243
877-506-NAFI

National Association of Hispanic Firefighters
2821 McKinney Avenue
Suite 7
Dallas, TX 75204
214-631-0025

National Association of State EMS Directors
201 Park Washington Court
Falls Church, VA 22046-4527
703-538-1799

National Association of State Fire Marshals (NASFM)
1319 F Street, NW
Suite 301
Washington, DC 20004
202-737-1226

National Collegiate EMS Foundation (NCEMSF)
PO Box 93
West Sand Lake, NY 12196
208-728-7342

National Council of State EMS Training Coordinators Inc (NCSEMSTC)
201 Park Washington Court
Falls Church, VA 22046
703-538-1794

National Emergency Management Association
PO Box 11910
Lexington, KY 40578
859-244-8000

National Emergency Medicine Association
306 West Joppa Road
Baltimore, MD 21204
410-494-0300

National Emergency Number Association (NENA)
4350 North Fairfax Drive
Suite 750
Arlington, VA 22203
800-332-3911

National Fire Academy Alumni Association
16825 South Seton Avenue
Emmittsburg, MD 21727
301-447-1000

National Fire Protection Association (NFPA)
1 Batterymarch Park
Quincy, MA 02169
617-770-3000

National Gang Crime Research Center
PO Box 990
Peotone, IL 60468
708-258-9111

National Institutes of Health (NIH)
9000 Rockville Pike
Bethesda, MD 20892
301-496-4000

National Organization for Victim Assistance (NOVA)
510 King Street
Suite 424
Alexandria, VA 22314
703-535-NOVA

National Propane Gas Association
1150 17th Street, NW
Suite 310
Washington, DC 20036
202-466-7200

National Registry of Emergency Medical Technicians (NREMT)
Rocco V. Morando Building
6610 Busch Blvd
PO Box 29233
Columbus, OH 43229
614-888-4484

National Study Center for Trauma and Emergency Medical Systems
701 West Pratt Street
Fifth Floor
Baltimore, MD 21201
410-328-5085

National Transportation Safety Board Communications Center
490 L'Enfant Plaza, SW
Washington, DC 20594
202-314-6000

National Volunteer Fire Council (NVFC)
1050 17th Street, NW
Suite 490
Washington, DC 20036
202-887-5700

Police and Firemen's Insurance Association
101 East 116th Street
Carmel, IN 46032
317-581-1913

React International Inc
5210 Auth Road #403
Suitland, MD 20746
301-316-2900

Society for Academic Emergency Medicine
901 North Washington Avenue
Lansing, MI 48906-5137
517-485-5484

Society of Critical Care Medicine
701 Lee Street
Suite 200
Des Plaines, IL 60016
847-827-6869

The Institution of Fire Engineers
United States of America Branch
4611 Strathblane Place
Alexandria, VA 22304
703-751-6416

United States Fire Administration
16825 South Seton Avenue
Emmitsburg, MD 21727
301-447-1000

Women in the Fire Service (WFS)
PO Box 5446
Madison, WI 53705
608-233-4768

Photo Credits

A

A.1 Courtesy of AAOS; A.2, A.4 © Jones and Bartlett Publishers. Courtesy of MIEMSS; A.3 Courtesy of NFPA; A.5 Courtesy of Survivair; A.6 Courtesy of AAOS

B

B.1 © Jones and Bartlett Publishers; B.2-B.5 © Jones and Bartlett Publishers. Courtesy of MIEMSS; B.6-B.8 Courtesy of NFPA

C

C.1, C.5, C.6, C.10, C.16, C.18, C.20 © Jones and Bartlett Publishers. Courtesy of MIEMSS; C.2, C.3, C.8, C.12, C.13, C.15 Courtesy of NFPA; C.4 Courtesy of www.aci-monitoring.com. Doug Beaulieu; C.7 Courtesy of Tyco Fire and Building Products; C.9 Courtesy of Draeger Safety, Inc.; C.11, C.14 © Jones and Bartlett Publishers; C.17, C.19 © Jones and Bartlett Publishers. Photographed by Philip Regan

D

D.1 © Bill Karrow, *Dunkirk Observer*/AP Photos; D.2 © Gregory Price, *Lewiston Sun Journal*/AP Photos; D.3 Courtesy of BRK Brands, Inc.; D.4 © Jones and Bartlett Publishers; D.5 Courtesy of Captain David Jackson, Saginaw Township Fire Department; D.6 © Jones and Bartlett Publishers. Courtesy of MIEMSS; D.7 © Jim Smalley

E

E.1 Spencer Grant/PhotoEdit; E.2, E.3, E.5-E.10 © Jones and Bartlett Publishers. Courtesy of MIEMSS; E.4 Courtesy of AAOS; E.11 Courtesy of Amerex Corporation; E.12 © Tonis Valing/ShutterStock, Inc.

F

F.1, F.6 © Jones and Bartlett Publishers. Photographed by Philip Regan; F.2, F.5, F.7, F.15, F.16 © Jones and Bartlett Publishers. Courtesy of MIEMSS; F.3, F.11-F.13 Courtesy of NFPA; F.4 © Jones and Bartlett Publishers; F.8 Courtesy of Captain David Jackson, Saginaw Township Fire Department; F.9 Courtesy of Charles B. Hughes/Unified Investigations & Sciences, Inc.; F.10 Courtesy of Anchor Industries, Inc.; F.14 © Jones and Bartlett Publishers

G

G.1 © Jones and Bartlett Publishers. Photographed by Philip Regan; G.2 Courtesy of NFPA; G.3 © Patrick Schneider, *The Charlotte Observer*/AP Photos

H

H.1, H.6, H.8, H.9 © Jones and Bartlett Publishers. Courtesy of MIEMSS; H.2 © Jones and Bartlett Publishers. Photographed by Philip Regan; H.3 Courtesy of NFPA; H.4 © Steve Allen/Brand X Pictures/Alamy Images; H.5 Courtesy of Firetronics Pte Ltd.; H.7 © Photodisc

I

I.1, I.9 © Jones and Bartlett Publishers; I.2, I.3 Courtesy of NFPA; I.4, I.5 © Dan Myers; I.6, I.8 © Jones and Bartlett Publishers. Courtesy of MIEMSS; I.7 Courtesy of Ralph G. Johnson (nyail.com/fsd)

L

L.1-L.5, L.7, L.8 © Jones and Bartlett Publishers. Courtesy of MIEMSS; L.6 Courtesy of Lakeland Industries, Inc.; L.9 © Keith D. Cullom

M

M.1, M.3, M.5, M.7 © Jones and Bartlett Publishers. Courtesy of MIEMSS; M.2 © Jones and Bartlett Publishers. Photographed by Philip Regan; M.4, M.9 © Jones and Bartlett Publishers; M.6 Courtesy of NFPA; M.8 Courtesy of Dennis Wetherhold, Jr.

N

N.1 © Jones and Bartlett Publishers. Courtesy of MIEMSS

O

O.1 © Jones and Bartlett Publishers; O.2 © Jones and Bartlett Publishers. Courtesy of MIEMSS

P

P.1, P.4-P.7, P.12-P.19 © Jones and Bartlett Publishers. Courtesy of MIEMSS; P.2, P.3, P.9, P.21 Courtesy of NFPA; P.8 Courtesy of Dennis Wetherhold, Jr.; P.10 Courtesy of AAOS; P.11 Courtesy of Duo-Safety Ladder Corporation; P.20 Courtesy of Captain David Jackson, Saginaw Township Fire Department; P.22 © Jones and Bartlett Publishers

R

R.1, R.5, R.7 Courtesy of NFPA; R.2 © Jeff Cooper, *Salina Journal*/AP Photos; R.3, R.4 © Jones and Bartlett Publishers. Courtesy of MIEMSS; R.6 Courtesy of AAOS

S

S.1, S.3, S.6, S.8 © Jones and Bartlett Publishers. Courtesy of MIEMSS; S.2 © Jones and Bartlett Publishers. Photographed by Philip Regan; S.4, S.13-S.15 © Jones and Bartlett Publishers; S.5 Courtesy of Super Vacuum Mfg. Co., Inc.; S.7, S.9 Courtesy of NFPA; S.10 Courtesy of Ralph G. Johnson (nyail.com/fsd); S.11 Courtesy of Duo-Safety Ladder Corporation; S.12 © 2003, Berta A. Daniels

T

T.1, T.2, T.5-T.7 © Jones and Bartlett Publishers. Courtesy of MIEMSS; T.3 © SHOUT/Alamy Images; T.4 Courtesy of Captain David Jackson, Saginaw Township Fire Department; T.8, T.9 Courtesy of Sandia National Laboratories

V

V.1, V.2, V.4 © Jones and Bartlett Publishers; V.3 Courtesy of Lakeland Industries, Inc.

W

W.1 Courtesy of USDA Forest Service; W.2 © Karen Wattenmaker Photography

DATE DUE

Hartness Library
Vermont Technical College
One Main St.
Randolph Center, VT 05061